Arthropods - New Advances and Perspectives

Edited by Vonnie D.C. Shields

Published in London, United Kingdom

Arthropods - New Advances and Perspectives
http://dx.doi.org/10.5772/intechopen.100916
Edited by Vonnie D.C. Shields

Contributors
Luciana Barboza Silva, Raimundo Henrique Ferreira Rodrigues, Thiago Ferreira Rodrigues, Maria Carolina Farias e Silva, Edivania de Araujo Lima, José Wellington Batista Lopes, Kenji Toyota, Taisen Iguchi, Yuta Sakae, Anil Chandra Neupane, Mikhail A. Ostrovskiy, Aleksander E. Dontsov, Jonathan Ibrahim, Dalyop Daniel Gyang, Amlan Das, Biswarup Mitra, Kewal Krishan Sharma, Thamilarasi Kandasamy, Ariane Flávia do Nascimento, Simone Magela Moreira, Bruna Macena Pereira de Souza, Muhammad Yasin, Amna Khan, Muhammad Anjum Aqueel, Muhammad Irfan Akram, Muhammad Aslam Farooqi, Hafiz Muhammad Bilal Yousuf, Muneba Noor, Aneeqa Maqsood, Max François Millien, Daphenide Saint-Louis, Daphnée Michel, Vonnie D.C. Shields

First published in London, United Kingdom, 2023 by IntechOpen
IntechOpen is the global imprint of INTECHOPEN LIMITED, registered in England and Wales, registration number: 11086078, 5 Princes Gate Court, London, SW7 2QJ, United Kingdom

British Library Cataloguing-in-Publication Data
A catalogue record for this book is available from the British Library

Additional hard and PDF copies can be obtained from orders@intechopen.com

Arthropods - New Advances and Perspectives
Edited by Vonnie D.C. Shields
p. cm.
Print ISBN 978-1-80355-612-3
Online ISBN 978-1-80355-613-0
eBook (PDF) ISBN 978-1-80355-614-7

For EU product safety concerns:
IN TECH d.o.o., Prolaz Marije Krucifikse Kozulić 3, 51000 Rijeka, Croatia,
info@intechopen.com or visit our website at intechopen.com.

Meet the editor

Vonnie D.C. Shields, Ph.D., is currently a full professor in the Biological Sciences Department and the associate dean of the Fisher College of Science and Mathematics, Towson University, USA. Dr. Shields' research explores gustatory, olfactory, and visual cues in insects. Her laboratory employs morphological, behavioral, and electrophysiological techniques to better understand sensory mechanisms by which larval and adult insects find host plants and detect plant-associated volatiles. Another line of research involves determining olfactory chemicals eliciting walking behavior. Dr. Shields received both a BS and Ph.D. from the University of Regina, Saskatchewan, Canada. A portion of her Ph.D. studies was carried out at the University of Alberta, Edmonton, Canada. After graduating, she accepted a research associate position to conduct postdoctoral studies at the Arizona Research Laboratories Division of Neurobiology, University of Arizona, USA, before she joined the faculty at Towson University where she rose through the ranks from assistant to full professor.

Contents

Preface

This book includes contributions on various topics pertaining to arthropods written by experts in their respective fields. It is an invaluable resource for entomologists, biologists, ecologists, zoologists, teachers, and students. The topics presented in this book are organized under four main sections: "Development", "Food Detection and Feeding Behavior", "Vector-Borne Diseases", and "Structure and Function of Vision".

In the first section, Chapter 1, "Larval Development of Non-Insect Arthropods: Metamorphosis and Sexual Differentiation", discusses the larval development of non-insect arthropods with respect to metamorphosis, sexual determination and differentiation, and the role of juvenile and molting hormones. More specifically, a comparison is made to insect metamorphosis, but emphasis is placed on decapod crustaceans and chelicerates. With respect to sex determination and differentiation, phenotypic differences that allow for sexual dimorphism of animal morphology, physiology, and behavior are discussed. The role of the DMRT gene is addressed in terms of its role in sex determination.

Chapter 2, "Spatiotemporal Dynamics of *Bemisia tabaci* MEAM1 (Hemiptera: Aleyrodidae) in Commercial Soybean Crops", covers spatiotemporal dynamics of crop pests in determining the colonization and dispersion patterns of insects using geostatistics. More specifically, it discusses devastating crop losses in soybeans attributed to whitefly populations and their monitoring and management.

In the second section, Chapter 3, "Food Detection and Feeding Behavior of Three Species of Household Cockroaches, *Blatella germanica* (L.), *Periplaneta americana* (L.), and *Supella longipalpa* (F.)" analyzes the importance of these pests. These insects have been blamed for producing/contributing to medical conditions such as allergies and asthma. In addition, they are vectors that transmit pathogenic bacteria. The author investigates the physiology of all three species with a focus on their circadian rhythm patterns, visual sensitivity, food-searching behavior, and attraction to selected foods. The development of various bait traps and the insects' resistance to them is covered as well.

Chapter 4, "Interaction Among the Multi-Trophic Lac Insect Complex of Flora and Fauna: Impact on Quantity and Quality of the Resin Secreted", covers a specialized group of phytosuccivorous insects that secrete resin of industrial importance. Factors affecting lac production and quality are discussed, as well as major and minor predators, parasites, and diseases capable of destroying lac crops. The effect of fungi on lac production and the interaction of lac insects with microbes are covered as well.

Chapter 5, "The Ability of Insects to Degrade Complex Synthetic Polymers", covers insect gut anatomy. It provides information about the synergy between gut microorganisms and the digestive enzymes of selected lepidoptera, coleoptera, and orthoptera in breaking down complex synthetic hydrocarbon polymers, such as polyurethane, polypropylene, polystyrene, and polyvinyl chloride.

Chapter 6, "Deterrents and Their Effects on the Feeding Behavior and Sensory Physiology of Insects", examines the gustatory system of lepidopterous insects. These insects have served as prominent models for carrying out neuroscience research. The chapter discusses the structural organization of taste organs as visualized by scanning and transmission electron microscopy. In addition, the author provides a discussion of the responses of selected lepidopterous larvae to secondary plant compounds, such as glucosinolates. More specifically, details are included as to how taste signals are recognized, coded, and processed by receptor cells housed in two specific sensory organs (sensilla), namely, the medial and lateral styloconic sensilla. These sensillas serve as the primary organs involved in feeding when the caterpillar feeds. The author examines the feeding behavior, hostplant preferences, and neurophysiological responses of sensory organs involved in peripheral gustatory coding.

In the third section, Chapter 7, "Vector-Borne Diseases and Climate Change", discusses various arthropod species (e.g., ticks, fleas, sandflies, mosquitoes, triatomine bugs, and blackflies) as vectors of numerous protozoan, bacterial, and virus insect-borne diseases, which result in Dengue fever, West Nile Virus, Lyme disease, malaria, and tick-borne diseases. The importance of climate change on vector-borne diseases and how it influences the survival, reproduction, abundance, and spatiotemporal distribution of vectors is explored. The rate of development and survival of pathogens within the vector-host is also considered.

Chapter 8, "Overview of the Main Species of Ticks and Animal and Human Tick-Related Diseases in the Caribbean, Particularly in Haiti", looks at the wide diversity of ticks and tick-borne diseases in humans and other animals in the Caribbean region. More specifically, the chapter focuses on the classification, morphology, and life cycle of ticks, as well as the transmission of tick-borne diseases. The authors explore the physical, geographical, and climatological characteristics and provide an analysis of epidemiological surveillance of pathogens.

Chapter 9, "Artrhopods: Prospect of Household Food Security", covers food inadequacy and the shortage of protein-rich foods in low-income households in developing countries and the exploitation of nutrient potentials acquired from using arthropods as food sources. It covers topics such as the nutritional composition of edible insects (i.e., protein, carbohydrate, mineral/vitamin, lipid, and secondary metabolites) as well as medicinal benefits, farming, and harvesting.

Chapter 10, "Capybara Ticks and the Urban Context of Spotted Fever in Brazil: An Overview", investigates how Capybara ticks play a detrimental role in amplifying Brazilian spotted fever among tick vector populations. The authors discuss how this situation leads to significant negative health impacts. Overall, the chapter discusses various aspects of Brazilian spotted fever, its urban occurrences and relationships between humans and other animal species, and the design of strategies and policies to protect the health of ecosystems.

In the fourth and final section, Chapter 11, "Ommochromes of the Compound Eye of Arthropods from the Insects and Crustaceans Classes: Physicochemical Properties and Antioxidant Activity", examines the main functions of ommochromes, their chemical structure and biosynthesis, and the physicochemical properties of isolated ommochromes using absorption and fluorescence spectroscopy and liquid chromatography. The antioxidant activity of ommochromes is included, as is a discussion of new pharmacological preparations for the prevention and treatment of ocular pathologies.

I wish to thank IntechOpen for initiating this book project and inviting me to serve as the editor. I would like to acknowledge Publishing Process Manager Romina Rovan Bakarčić for guiding me through the process. I would like to thank all the authors who contributed to this book for their hard work in submitting and editing their contributions. I would also like to thank an anonymous reviewer for revising the chapter that I contributed. Finally, I thank my husband Dr. Thomas Heinbockeland, and our son Torben Heinbockel for their patience and understanding when I was working on this book project.

Dr. Vonnie D.C. Shields
Associate Dean,
Jess and Mildred Fisher College of Science and Mathematics,
Towson University,
Towson, MD, USA

Professor,
Biological Sciences Department,
Towson University,
Towson, MD, USA

Section 1

Development

Chapter 1

Larval Development of Non-Insect Arthropods: Metamorphosis and Sexual Differentiation

Kenji Toyota, Yuta Sakae and Taisen Iguchi

Abstract

In insects, metamorphosis is one of the most important research topics. Their drastic morphological and physiological changes from larvae to pupae, and then to adults, have fascinated many people. These changing life history patterns are tightly regulated by two endocrine systems, the ecdysteroids (molting hormones) and the juvenile hormones. Metamorphosis is also the most universal phenomenon in non-insect arthropods (especially crustaceans). Additionally, as dwarf males (e.g., barnacle crustaceans) show distinct sexual dimorphism during the larval developmental stage, larval development and sexual differentiation are also intimately associated. Our knowledge of endocrinology and gene cascades underlying metamorphosis and sexual differentiation in non-insect arthropods is rudimentary at best and relies heavily on well-studied insect models. Advances in newly developed applications, omics technologies and gene-targeting, are expected to lead to explorative molecular studies that reveal components and pathways unique to non-insect arthropods. This chapter reconciles known components of metamorphosis and sexual differentiation in non-insect arthropods and reflects on our findings in insects to outline future research.

Keywords: metamorphosis, ecdysteroid, juvenile hormone, sex determination, *doublesex*

1. Introduction

Arthropods are among the best-known animals on the earth and have fascinated many people, including researchers. Extant arthropods can be classified into four sub-families: Chelicerata, Myriapoda, Crustacea, and Hexapoda. Although it is now widely accepted that Crustacea and Hexapoda are integrated as Pancrustacea, Crustacea and Hexapoda (insects) are used to focus on non-insect arthropods in this chapter. Their phylogenetic relationship has been debated for many years. However, recent progress in next-generation sequencing has provided their exact position (**Figure 1**) [1–3]. Genetics and developmental biology using model insects such as fruit fly *Drosophila melanogaster*, red flour beetles *Tribolium castaneum*, and silkworm *Bombyx mori*, have long been a driving force in not only basic biology but also a wide range of sciences including medical, agricultural, and so on. On the other hand, for non-insect arthro-pods, there is a long history of physiological knowledge of some crustaceans (especially

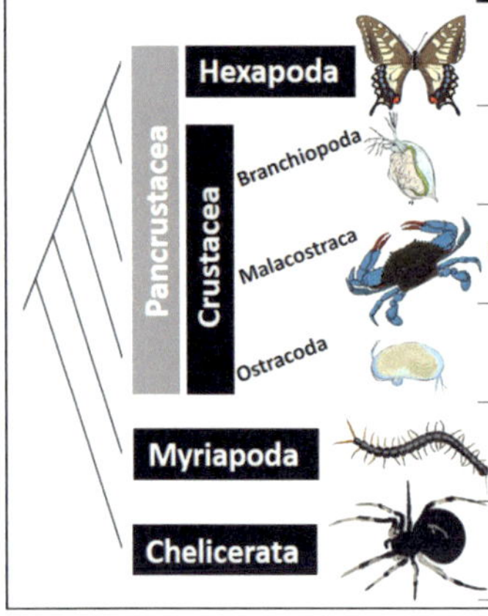

	Innate JH / ecdysteroids	Synthesizing organs	Synthesizing enzymes in genome	Receptors in genome	Sex hormone	Regulation of dsx
Hexapoda	JH III etc. / 20E	Corpora allata / Prothorathic gland	Yes / Yes	Yes / Yes	?	Sex-specific splicing
Branchiopoda	MF / 20E	? / Gut	Yes / Yes	Yes / Yes	MF	Male-biased expression
Malacostraca	MF etc. / 20E	Mandibular organ / Y-organ	Yes / Yes	Yes / Yes	IAG, CFSH	Female- or male-biased expression
Ostracoda	? / ?	? / ?	? / ?	? / ?	?	?
Myriapoda	? / ?	? / ?	Yes / Yes	Yes / Yes	?	?
Chelicerata	? / 20E?	? / ?	Yes / Yes	Yes / Yes	?	Male-specific expression?

Figure 1.
Phylogenetic tree of extant arthropods. The branching pattern is constructed based on previous studies [1–4] with the exclusion of some clades for clarity. Crustacea are not monophyletic and include Ostracoda, malacostraca, decapods, and Branchiopoda. Detection of synthesizing enzyme genes in the genomes is indicated by the presence of putative juvenile hormone acid O-methyltransferase (JHAMT) or farnesoic acid O-methyltransferase (FAMeT) for JH, and of putative Halloween genes for ecdysteroids, respectively. Likewise, in terms of these receptor genes in the genome, it is indicated by the presence of methoprene tolerant (met) orthologs for JH and of ecdysone receptor (EcR) for ecdysteroids, respectively.

decapods) that are important for fisheries and aquacultures and some mites that are problematic as agricultural pests, but molecular insights are still limited due to lack of genome information and reverse genetic approaches. In the last decade, the research environment surrounding biology has drastically improved with the advances of sequencing, imaging, handling of large-scale data (bioinformatics), and new-generation genome modification technologies such as genome editing. Based on these developments, many findings on embryogenesis, larval metamorphosis, sex determination, and sexual differentiation have been reported in many non-insect arthropods. In this chapter, we will provide an overview of the knowledge of canonical metamorphosis and sex determination of insects, and the attempt to compare them with non-insect arthropods, particularly branchiopod and decapod crustaceans and spider chelicerates.

2. Larval metamorphosis

Tadpoles developing into frogs, and insect larvae or pupae into adults, are classical examples of metamorphosis, and biologists have long been fascinated with these dramatic and obviously spontaneous transformations [5–7]. Previous studies showed that such morphological alterations were accompanied by major changes in the chemical composition and biochemical function of almost all larval tissues. In the last century, the discovery that metamorphosis is centrally regulated by endocrine systems has allowed us to understand how these post-embryonic developmental and physiological processes are brought about and controlled [7, 8]. This section outlines the endocrine and molecular mechanisms of canonical complete metamorphosis (holometaboly) in insects and then overviews current research findings on larval metamorphosis in the decapod crustaceans and arachnid chelicerates.

2.1 Insects with complete metamorphosis

Holometabolous insects go through a series of discrete stages (larva, pupa, and adult) that hardly resemble one another, but are finely adapted to specific roles in their life cycle. Juvenile hormone (JH) and ecdysteroids (the representative active form is 20-hydroxyecdysone: 20E) are the two key endocrine factors that together coordinate molting and metamorphosis (**Figure 2**) [7, 9–11].

JHs are a family of acyclic sesquiterpenoids that are involved in a range of physiological processes in insects such as not only metamorphosis, but also ovarian development, reproductive behavior, and various types of phenotypic plasticity including caste differentiation in social insects and weapon traits development in beetles [12, 13]. In terms of metamorphosis, JH acts as the best-known anti-metamorphic hormone which prevents larvae/nymphs from undergoing precocious metamorphosis and is often referred to as the *status quo* hormone [14]. Its representative *status quo* action is associated with a surge of ecdysteroids that triggers molting [15]. At the beginning of the ecdysteroids surge, the presence or absence of JH decides whether the molt will be a *status quo* molt that repeats the same stage (larva-larva) or a molt with metamorphosis (larva-pupa or pupa-adult) [16].

More than half a century of arthropod endocrinology research has revealed that molting is precisely regulated by complex multiple hormone systems, and ecdysteroids are a key hormone mediating a variety of physiological and behavioral changes that are essential for molting and metamorphosis [7, 10, 17]. Their primary function is to induce molting and they serve this function throughout the arthropods [10, 18]. In insects, circulating ecdysteroids typically come from the prothoracic glands. The prothoracic glands secrete ecdysone which is then further converted to 20E in peripheral tissues [19]. The functions of ecdysteroids in the control of insect metamorphosis have striking parallels with those of the thyroid hormones in directing the metamorphosis of amphibians [5, 6].

2.2 Larval metamorphosis in decapod crustaceans

Recent molecular phylogeny has supported the theory that the Crustacea clade is not monophyletic, but is divided into at least three extant clades (Ostracoda, Malacostraca, and Branchiopoda) (**Figure 1**) [4]. As aforementioned, both Crustacea and Hexapoda form a new clade known as Pancrustacea [1–3]. This theory spurs the notion that a comparative analysis between crustaceans and insects is essential to

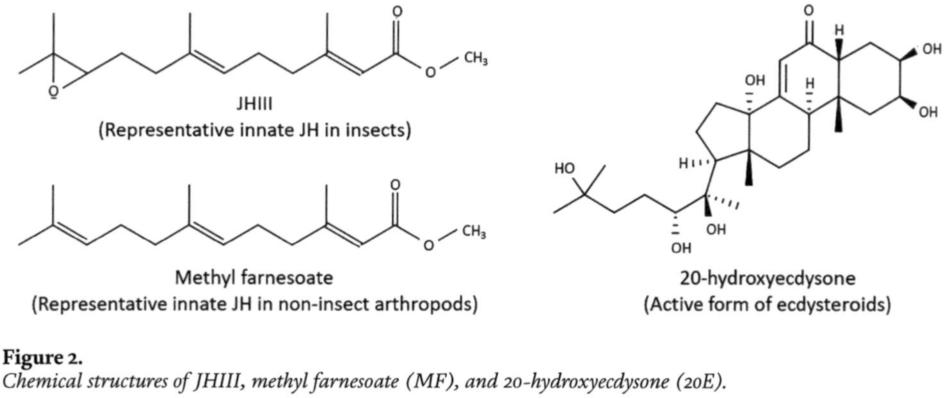

Figure 2.
Chemical structures of JHIII, methyl farnesoate (MF), and 20-hydroxyecdysone (20E).

understanding the evolutionary origins of various traits considered unique to insects. Indeed, crustaceans and insects share various basic traits, such as endocrine-driven developmental and reproductive processes, which are regulated primarily by JHs and ecdysteroids. In non-insect Arthropods, methyl farnesoate (MF) is thought to be the equivalent of JH in insects. In insects, MF is finally converted to JH III (active JH form in insects) by CYP15A1, except for the Lepidoptera [20]. However, CYP15A1 orthologs have never been found in non-insect Arthropoda [21, 22], indicating that the CYP15A1 gene acquisition might have been an important event enabling JH biosynthesis in insects [20]. In fact, detection of insect-type JH (e.g., JH III) has not been reported in crustaceans; instead, MF has been widely regarded as the functional crustacean JH since its discovery in various crustaceans [8, 23, 24]. Research on crustacean larval metamorphosis and endocrine pathways such as MF and ecdysteroids has been vigorously pursued in the fisheries-important decapod order (crabs and shrimps). Production of MF and ecdysteroids in decapod crustaceans is thought to take place in the mandibular organ and Y-organ, the glands unique to Malacostraca crustaceans which are considered to be functionally analogous to the insect corpus allatum and prothorathic glands [21]. Unlike insects, the synthesis of MF and ecdysteroids in decapod crustaceans is inhibitory regulated by mandibular organ-inhibiting hormones (MOIHs) and molt-inhibiting hormones (MIHs), cryptic members of the crustacean hyperglycemic hormone (CHH)-like neuropeptides secreted from the X-organ/sinus gland complex in the eyestalk [25].

In the marine decapod species, each larva differs from conspecific juveniles and adults based on morphological, ecological, behavioral, physiological, and/or other biological traits. Moreover, the juveniles and adults are benthic feeders or predators, whereas larvae grow in the pelagic environment, preying on phyto- and/or zoo-plankters [26]. In addition, those larval forms are so different from the conspecific adults that some of them have been described as distinct species, and many larval names were originally genus names, for example, nauplius, cypris, zoea, megalopa, mysis, nisto, puerurus, and phyllosoma in decapod species [26]. This disorganization of larval names causes confusion in advancing comparative developmental and physiological biology in decapods; however, no unified nomenclature has yet been defined. To resolve this problem, clarification of the endocrine regulation of larval metamorphosis and the associated gene regulatory network among decapod species would facilitate interspecific comparative analysis and advance our understanding of this phenomenon, as is the case with insects. Although several studies have challenged the current understanding of the effects of MF and 20E on larval metamorphosis, no consistent results have been obtained, as follows.

Our group demonstrated that treatment of either MF or 20E induced high mortality caused by disruption of molting-associated metamorphosis in the kuruma prawn *Marsupenaeus japonicus* (larval stages: nauplius, zoea, mysis, post larva, and juvenile; **Figure 3**) [27]. Likewise, treatment of MF to the larvae of the freshwater prawn *Macrobrachium rosenbergii* resulted in an inhibitory effect on metamorphosis [28]. On the other hand, the administration of MF has been shown to accelerate metamorphosis in late-stage shrimp and barnacle larvae [23, 29–31]. These data suggest that the role of MF in regulating decapod metamorphosis is still unclear. As a distinct approach, eyestalk ablation of intermediate larvae led to failure or impedance of metamorphosis to the megalopa stage in several crabs *Rhithropanopeus harrisii*, *Callinectes sapidus* and *Portunus trituberculatus*, and American lobster *Homarus americanus* [32–34]. Later, the impact of eyestalk ablation on metamorphosis was demonstrated to be accompanied by a rise in circulating MF concentrations,

Figure 3.
Larval developmental staging (nauplius, zoea, and mysis) and adult (dorsal and frontal views) of kuruma prawn. Each scale bar indicates 100 μm.

confirming that these observations were potentially the result of inhibition by MF [35]. Additionally, as for the new mystery, our previous work demonstrated that nauplius larvae of kuruma prawn have higher tolerance against the lethal effect of MF and 20E treatment than other later stages (e.g., zoea and mysis), suggesting that there are different endocrine cassettes regulating transition from nauplius to zoea and later metamorphosis [27]. Similarly, larval transcriptomics during metamorphosis in spiny lobster *Sagmariasus verreauxi* has suggested that the puerulus (megalopa)-juvenile metamorphosis is regulated by the MF signaling pathway, whereas the phyllosoma (zoea)-puerulus (megalopa) metamorphosis is preceded by sustained the expression level of FAMeT, which encodes the enzyme synthesizing MF [36]. These data could suggest that MF is not involved in early-stage (nauplius in the kuruma prawn and phyllosoma in spiny lobster) metamorphosis through the conventional inhibitory mechanism.

2.3 Larval metamorphosis in chelicerates

Chelicerates are classified into a large and ancestral sister group of arthropods (**Figure 1**). Chelicerates have generally been considered to be ametabolous because the larvae display a very similar morphology to adults (**Figure 4**). However, detailed morphological observations in several chelicerates cast doubt that the chelicerates are "ametabolous."

Observations with a scanning electron microscope show that scorpion larvae have incomplete mechanoreceptors and chemoreceptors [37]. Larvae differ in the morphology of the tip of tarsi and aculeus compared to that of nymphs and adults [37, 38]. The scorpion larvae ride on the mother's back after the delivery, at which point the

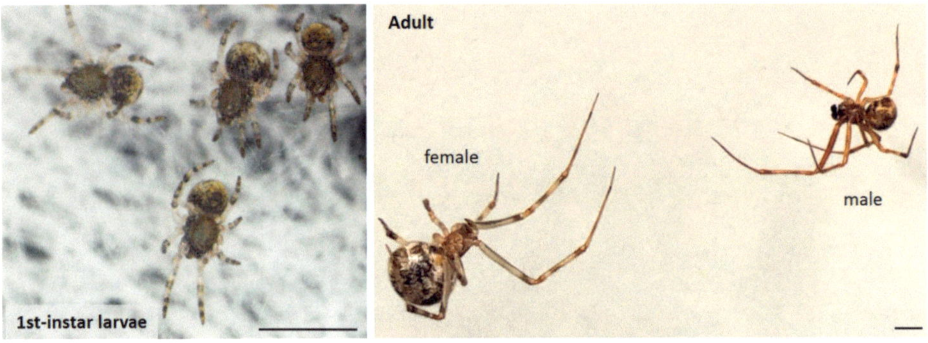

Figure 4.
The appearance of larvae and adult in Parasteatoda tepidariorum. Scale bars indicate 1.0 mm.

structure at the tip of the larva's tarsi works, the tip of the tarsi functions like a sucker, before developing into a claw after molting [37, 38]. In addition, the exoskeleton of adult scorpions fluoresces when exposed to UV light, whereas the exoskeleton of first-instar larvae does not fluoresce [39, 40]. In ticks, the genital primordium opens to the exoskeleton through multiple molting [41]. The fourth leg, not observed in first-instar mite larvae, appears in molted nymphs [42]. In spiders (*Parasteatoda tepidariorum*), the development of the pedipalp, which is the male copulatory organ, progresses at the sub-adult stage similarly to the metamorphic manner of insects and is completed through molting to the adult [43]. Thus, morphological observations of larvae and sub-adults in scorpions, ticks, and spiders suggest that chelicerates exhibit hemimetaboly rather than ametaboly.

In insects and crustaceans, it is demonstrated that ecdysteroids and MF regulate metamorphosis, but the molecular mechanism of metamorphosis in chelicerates is unknown [8, 12, 44–46]. In chelicerates, 20E has been detected in ticks, scorpions, horseshoe crabs, and spiders [47–52]. The ecdysone receptor (EcR) and retinoid X receptor (RXR), which act as 20E receptors, have also been identified in ticks, scorpions, and spiders [53–59]. In fact, the administration of 20E demonstrates to induce molting of ticks, horseshoe crabs, and spiders [60–63]. Although there have been no reports of identification of sesquiterpenoid hormones in Chelicerata, genomic and transcriptomic data suggest that ticks and spiders may have precursors of JH, MF [64, 65]. Therefore, it is possible that ecdysteroids and sesquiterpenoid hormones cause molting and associated metamorphosis in chelicerates, despite this not being direct evidence.

3. Sex determination and differentiation

Sex determination is the most fundamental developmental process that establishes sexually dimorphic traits. In many animals, males and females have distinct sex-related characteristics such as body size, ornamentation, and color [66]. These extreme phenotypic differences make sexual dimorphism one of the most interesting aspects of animal morphology, physiology, and behavior. This outstanding diversity of sexually dimorphic traits is reflected in the underlying molecular mechanisms by a series of systems ranging from sex-specific gonadal steroid hormones sealing sexual fate in mammals and other vertebrates [67], to cell-autonomous sex-specific splicing loops that maintain the sexual state in holometabolous insects [68, 69].

The DMRT gene, which is conserved in metazoans, is an important transcriptional factor that has a key role in sex determination [70–72]. Four DMRT genes (*doublesex, dmrt11E, dmrt93B, dmrt99B*) have been found in insects, and *doublesex* (*dsx*) is known as a conserved master switch regulating the development of sexual dimorphic traits [72, 73]. *Dsx* contributes to insect sex determination via sex-specific splicing cascades [68, 72]. All insect species have cell-autonomous sex determination mechanisms, in which the sex-determining cascade operates on a cell-to-cell basis [74].

Focusing on the mechanisms inducing the sex-specific traits by DSX function, there are major differences between insects and non-insect arthropods. As described above, in insects, sex-specific splicing of *dsx* makes sexual dimorphic traits, although the upstream signals are extremely variable in the insect order [75]. On the other hand, previous studies using the water flea, which is a well-studied species in the branchiopod crustaceans, have revealed that its *dsx* has no sex-specific splicing isoforms and apparently shows the male-biased expression pattern resulting in the promotion of masculinization [76–79]. As described below, there is no sex-specific splicing of *dsx* in non-insect arthropods, and the mechanism of sexual differentiation by male-biased expression is conserved.

3.1 Decapod crustaceans as a treasury box of diverse sex differentiation pathways

In general, sex determination and sexual differentiation in arthropods are cell-autonomous in manner, whereas in most mammals and other vertebrates, sexual identity is unified throughout the body by sex steroids (estrogens and androgens) secreted by the gonads. Interestingly, only Malacostraca crustaceans including decapods exceptionally have a cell-nonautonomous sexual differentiation manner, and unlike gonad-dependent endocrine regulation in vertebrates, have male-specific endocrine glands known as the androgenic glands (AG), which are located on the terminal of the vas deferens [80]. Physiological roles of the AGs have been historically revealed to play a key function in male sexual differentiation, by AG ablation and implantation in the amphipod *Orchestia gammarella* [80]. Other studies have been conducted in the woodlouse *Armadillidium vulgare* using AG implantation [81], AG ablation [82], and injections of AG extracts [83]. Then, the androgenic gland hormone has been purified and identified [84, 85]. After that, physiological roles of AGH have further been demonstrated in decapod species using, for instance, AG implantation in the red claw crayfish *Cherax quadricarinatus*, and the marbled crayfish *Procambarus fallax f. virginalis* [86]. Moreover, our group has recently succeeded in chemically synthesizing *P. fallax f. virginalis* insulin-like androgenic gland factor (IAG) [87]. Although injection of the synthetic IAG to female crayfish did not induce masculinization on the external morphology, this treatment apparently suppressed oocyte maturation *in vivo* [87], suggesting that its IAG has a pivotal role in the suppression of female secondary sex characteristics. Several attempts to understand the regulatory mechanism of IAG expression have shown that male eyestalk ablation causes AG hypertrophy and hyperplasia [88, 89]. There is a unique developmental axis defined as the X-organ/sinus gland/neuroendocrine complex (XO-SG)-AG-testis axis. Although the fine details are controversial, it has been demonstrated that IAG interacts with its binding protein and receptor to activate downstream pathways [90–92].

Several studies have demonstrated that the *dsx* regulates the IAG expression. For example, the *dsx* is predominantly expressed in the testis, and its expression levels gradually increase with larval development in the Chinese shrimp *Fenneropenaeus chinensis*. Its knockdown resulted in suppression of *IAG* expression, suggesting that *dsx* promotes male sexual differentiation via IAG signaling [93]. While, the *dsx* mainly expresses in the ovary, and its knockdown increased *IAG* expression in the red claw crayfish *C. quadricarinatus*, implying that *Cqdsx* is involved in female sexual differentiation [94]. Both *dsx* genes have no sex-specific splicing forms, therefore, there is male- or female-biased expression to promote sexual differentiation pathways.

In addition to IAG, the crustacean female sex hormone (CFSH) was discovered as a novel eyestalk-derived neuropeptide that induces the development of secondary female characteristics in two crab species *C. sapidus* and *Carcinus maenas* [95]. CFSH is synthesized and secreted in the X-organ/sinus gland complex in the eyestalks.

Currently, its homolog has been successfully identified in several other decapod species such as the swimming crab *P. trituberculatus* [96], the Chinese mitten crab *Eriocheir sinensis* [97], the green shore crab *C. maenas* [98] and the mud crab *Scylla paramamosain* [99–101], the kuruma prawn *M. japonicus* [102], the Pacific white shrimp *Litopenaeus vannamei* [97], the banana shrimp *Fenneropenaeus merguiensis* [103], the Antarctic shrimp *Chorismus antarcticus* [104], the Eastern rock lobster *S. verreauxi* [105], the giant freshwater prawn *M. rosenbergii* [97, 106, 107], the peppermint shrimp *Lysmata vittata* [108, 109], the red swamp crayfish *Procambarus clarkii* [96], and the Australian crayfish *C. quadricarinatus* [110]. RNA interference of *CFSH* caused the anomalous development of female reproductive characteristics including ovigerous setae, gonopores, and extended parental brood care in *Callinectes sapidus* [95], and the formation of gonopores in juvenile stages in the mud crab [100]. These reports indicate that the CFSH plays a pivotal role in the development of female-specific reproductive characteristics. On the other hand, it has been discovered that CFSH expression can be detected in the eyestalks of both females and males in the kuruma prawn [111] and several crab species [95, 99, 100], and moreover, two distinct CFSH subtypes have been identified as eyestalk- and ovary-types in the kuruma prawn [111]. Based on immunohistochemistry and *in situ* hybridization analyses of both CFSH subtypes, the ovary-type is predominantly expressed in oogonia and previtellogenic oocytes during vitellogenesis. These data suggest that ovary-type CFSH may take part in reproductive processes, although the differences in physiological function between both subtypes are still unclear. Besides, in the Australian crayfish, CFSH expression has been detected in the central nervous system, antennal gland, and gut [110]. Taken together, accumulating CFSH studies indicate that it might regulate female secondary reproductive phenotypes in some crab species, and is not a female-specific hormone in other decapod species. Although a new potential function of CFSH has been reported to be its involvement in growth [108], more comparative analysis will be required for comprehensively understanding its physiological roles.

Some recent studies have demonstrated the crosstalk between CFSH and IAG to facilitate sexual differentiating processes. In the mud crab *S. paramamosain*, it has been demonstrated that CFSH plays a pivotal role in the development of female reproductive traits and suppresses the *IAG* expression in AG *in vitro* [99]. Furthermore, the transcriptional relationship of *CFSH* to *IAG* expression has also been demonstrated with respect to the involvement of signal transducers and activators of the transcription-binding site [100]. Additionally, it has recently demonstrated that feedback regulation of both IAG and CFSH in peppermint shrimp *L. vittata*, a species that possesses a protandric simultaneous hermaphroditism reproductive system [108, 109]. To date, the CFSH receptor has not been identified. Further studies on the CFSH receptor and its downstream signaling pathways are necessary to understand the mechanisms underlying endocrine crosstalk among CFSH, IAG and *dsx* in sexual differentiation of decapods.

3.2 Sex determination and differentiation in chelicerates

Since chelicerates is an ancestral sister group among arthropods, it is an important group for considering the evolutionary diversity of sex determination and sexual differentiation mechanisms including arthropods and vertebrates. Among chelicerates, spiders display a particularly clear morphological sexual dimorphism, females are 3–14 times larger than males and, in some species, females are 75.2 times heavier than males [112, 113]. In addition, several species of males, such as the banksia peacock spider, show a brilliant appearance like a peacock male and perform the mating dances [114, 115]. While studies on morphological and behavioral sexual dimorphism have been proceeding, studies on sex determination and sexual differentiation are completely unclear. Not only in spiders but also in other chelicerates, has research on sex determination and sexual differentiation remained almost untouched. However, recent studies have begun to find clues to the mechanism of sex determination and/or sexual differentiation in Chelicerata, referring to the sex determination cascade of insects.

In tick (*Metaseiulus occidentalis*), a comparative analysis of genomic sequences of insects and water fleas identified *dsx* (*MoccDsx1* and *MoccDsx2*), *dmrt11E* (*MoccDmrt11E*), and *dmrt99B* (*MoccDmrt99B*) [116, 117]. Quantitative RT-PCR in adult ticks indicates that *MoccDsx1* and *MoccDsx2* are highly expressed in males [116, 117]. Furthermore, the transcripts of *MoccDsx1* and *MoccDsx2* in adult males and females exhibit the same size, suggesting that they are not subject to sex-specific splicing [116, 117]. However, it is still unclear whether they actually exhibit transcriptional patterns and levels similar to adults at the sex determination point.

Spiders and scorpions experienced whole-genome duplication (WGD) after diverging from other chelicerates [118]. Seven *dsx*-like genes are found in the genome sequence of the spider (*P. tepidariorum*) [119]. Probably, it is expected that one of eight *dsx*-like genes in the ancestor disappeared after WGD and so the total became seven. The *dsx*-like genes of *P. tepidariorum* are classified by comparative analysis of gene and amino acid sequences of *dsx* in fly (*Drosophila melanogaster*) and water flea (*Daphnia magna*): *dsx* (*PtDsx1*), *dmrt11E* (*PtDsx2*), *dmrt93B* (*PtDsx3*), *dmrt99B* (*PtDsx1A*, *PtDsxA2* and *PtDsxA2-like*), other (*PtLOC107443841* and *PtLOC110283461*) [119]. A high transcript level of *dsx* (*PtDsx1*) and *dmrt99B* (*PtDsxA2-like*) in adult males and female bias expression of *dmrt93B* (*PtDsx3*) is detected [119]. While the sequence analysis of the cloned *dsx*-like genes cDNA detects splicing variants in *dsx* (*PtDsx1*) and *dmrt11E* (*PtDsx2*), there is no sexual difference in their expression. In addition, whole-mount *in situ* hybridization using embryos cannot provide information about the sex difference of each gene due to the lack of genetic sex markers in *P. tepidariorum*. Therefore, the results of expression analysis in sub-adults and adults of ticks and spiders suggest that sex determination in chelicerates might be caused by high expression of *dsx* in males, the same as in crustaceans.

4. Conclusions and future directions

This chapter focuses on larval metamorphosis, sex determination, and sexual differentiation in non-insect arthropods, especially in decapod crustaceans and spider chelicerates. Insects have long been the frontrunners in the study of these phenomena in arthropods, however, new emerging methods such as next-generation sequencing, 3D and high-resolution imaging techniques [120, 121], and genome editing methods [122–125] are opening the door for every non-model species. It is of great benefit to the wealth of available knowledge of insects. Indeed, although this chapter referred primarily to holometabolous insects, the latest work has revealed that some hemimetabolous insects such as termites do not have the sex-specific splicing isoforms of *dsx* observed in decapods and chelicerates [126–128]. Thus, new discoveries that overturn established theories are being found in insects, and it is expected that the developmental and physiological biology of non-insect arthropods will advance dramatically in the near future.

Although not described extensively in this chapter, the research environment is paving the way for Myriapoda species using the centipede *Strigamia maritima* as a model [129]. In 2014, the genome of this species was also released, revealing that the biosynthesis and receptor systems for JH and ecdysteroids are conserved (**Figure 1**) [129, 130]. The repertoire of DM domain-containing genes, including *dsx*, has also been reported, and the regulatory mechanism of *dsx* expression is expected to be elucidated in the near future.

To date, chelicerates have generally been considered ametabolous because their offspring display similar morphology to adults. However, detailed observations of several species revealed the morphological changes associated with molting. These results suggest that we need to change our perception to chelicerates as being hemimetabolous organisms.

Chelicerates tend to only be seen as pest organisms because they are a source of allergies and have venom. However, in recent years, among the chelicerates, spiders, and scorpions have begun to be emphasized as material sources and models that can play an active role in various industries. Spiders spin up to seven types of thread called "spider silk" [131–133]. It had been used in sutures and fishing lines in ancient times due to its lightweight, strong and extensible properties [131, 133, 134]. In addition to these, spider silk and its constituent spidroins are currently being considered for application in adhesives, cosmetics, humidity sensors, and the aerospace industry [132, 135, 136]. It is also attracting attention as a biomaterial for biomedical applications (artificial blood vessels, matrigels, porous sponges, and microcapsules) due to its high cell compatibility, low immunogenicity, and slow *in vivo* degradability [132, 137, 138]. In fact, spider silk is used as a guiding material and scaffolding for transplanted cells as demonstrated in a preclinical study for regenerative treatment of bone, skin, myocardium, and nerve [139, 140]. Spider silk, which is useful in various industries, has recently been found to have sex differences in composition [114]. If either male or female spider silk may be more suitable for the desired application, the development of a technique that can control the sex of the spider in order to increase the production of spider silk of either sex may be expected. Other benefits of developing techniques to control the sex of spiders and other chelicerates are the application of spider and scorpion venoms to pharmacology, medical science, and the development of agricultural pesticides. Spider and scorpion venom is a complex cocktail containing a lot of mineral salts, small organic molecules, and small polypeptides [141–143]. This cocktail contains unknown molecules that have a biomolecular activity and is of great interest. In fact, some poisons contribute to the development of pharmacologic tools (e.g., PcTx1; tool for elucidating the roles of acid-sensing ion channels and related pathologies), clinical trials (e.g., GsMTx4; for the treatment of Duchenne muscular dystrophy, and Chlorotoxin; an anticancer drug for neuroectodermal tumors), and pesticides (e.g., GS-ω/κ-HXTX-Hv1a; a bioinsecticide for aphids, spider mites, spotted-winged drosophila, thrips, and whiteflies recognized as major greenhouse pests: Spear® T (Vestaron Corporation, Durham, NC, USA)) [143, 144]. In recent years, it has been reported that spider and scorpion venom components also exhibit sex differences [145–149]. This suggests that the development of control techniques for chelicerates sex may be useful for efficient venom constituent identification and increased production. However, most sex-related research, including sex determination and sexual differentiation, remains unclear in chelicerates. This is due to a lack of analytical tools for late-stage embryos and larvae, including the markers for the typing of genetic sex. We hope that we will overcome these barriers in the future and deepen our understanding of sex in chelicerates, which is important from a biological, evolutionary, and industrial perspective.

Acknowledgements

The authors would like to thank Dr. Mike Roberts, Independent Consultants, UK for his critical readings of this manuscript.

Conflict of interest

The authors declare no conflicts of interest associated with this manuscript.

Author details

Kenji Toyota[1,2,3]*, Yuta Sakae[4,5] and Taisen Iguchi[6]

1 Niigata University, Marine Biological Station, Sado Center for Ecological Sustainability, Sado, Niigata, Japan

2 Faculty of Science, Department of Biological Sciences, Kanagawa University, Hiratsuka, Kanagawa, Japan

3 Faculty of Industrial Science and Technology, Department of Biological Science and Technology, Tokyo University of Science, Katsushika-ku, Tokyo, Japan

4 Department of Integrated Biosciences, The University of Tokyo, Graduate School of Frontier Sciences, Kashiwa, Chiba, Japan

5 Division of Translational Genomics, Exploratory Oncology Research and Clinical Trial Center, National Cancer Center, Kashiwa, Chiba, Japan

6 Yokohama City University, Graduate School of Nanobioscience, Yokohama, Kanagawa, Japan

*Address all correspondence to: megabass0719@yahoo.co.jp

References

[1] Schwentner M, Combosch DJ, Pakes Nelson J, Giribet G. A phylogenomic solution to the origin of insects by resolving crustacean-hexapod relationships. Current Biology. 2017;**27**:1818-1824.e5. DOI: 10.1016/j.cub.2017.05.040

[2] von Reumont BM, Jenner RA, Wills MA, Dell'ampio E, Pass G, Ebersberger I, et al. Pancrustacean phylogeny in the light of new phylogenomic data: Support for Remipedia as the possible sister group of Hexapoda. Molecular Biology and Evolution. 2012;**29**:1031-1045. DOI: 10.1093/molbev/msr270

[3] Giribet G, Edgecombe GD. The phylogeny and evolutionary history of arthropods. Current Biology. 2019;**29**:592-602. DOI: 10.1016/j.cub.2019.04.057

[4] Oakley TH, Wolfe JM, Lindgren AR, Zaharoff AK. Phylotranscriptomics to bring the understudied into the fold: Monophyletic Ostracoda, fossil placement, and pancrustacean phylogeny. Molecular Biology and Evolution. 2013;**30**:215-233. DOI: 10.1093/molbev/mss216

[5] Brown DD, Cai L. Amphibian metamorphosis. Developmental Biology. 2007;**306**:20-33. DOI: 10.1016/j.ydbio.2007.03.021

[6] Tata JR. Autoinduction of nuclear hormone receptors during metamorphosis and its significance. Insect Biochemistry and Molecular Biology. 2000;**30**:645-651. DOI: 10.1016/s0965-1748(00)00035-7

[7] Truman JW. The evolution of insect metamorphosis. Current Biology. 2019;**29**:1252-1268. DOI: 10.1016/j.cub.2019.10.009

[8] Cheong SPS, Huang J, Bendena WG, Tobe SS, Hui JHL. Evolution of ecdysis and metamorphosis in arthropods: The rise of regulation of juvenile hormone. Integrative and Comparative Biology. 2015;**55**:878-890. DOI: 10.1093/icb/icv066

[9] Ando H, Ukena K, Nagata S, editors. Handbook of Hormones Comparative Endocrinology for Basic and Clinical Research 2nd Edition. Volume 1 and Volume 2. Academic Press; 2021. p. 1113

[10] Song Y, Villeneuve DL, Toyota K, Iguchi T, Tollefsen KE. Ecdysone receptor agonism leading to lethal molting disruption in arthropods: Review and adverse outcome pathway development. Environmental Science & Technology. 2017;**51**:4142-4157. DOI: 10.1021/acs.est.7b00480

[11] Toyota K, Watanabe H, Hirano M, Abe R, Miyakawa H, Song Y, et al. Juvenile hormone synthesis and signaling disruption triggering male offspring induction and population decline in cladocerans (water flea): Review and adverse outcome pathway development. Aquatic Toxicology. 2022;**243**:106058. DOI: 10.1016/j.aquatox.2021.106058

[12] Nijhout HF. Insect Hormones. Princeton: Princeton University Press; 1994. p. 388

[13] Gotoh H, Cornette R, Koshikawa S, Okada Y, Lavine LC, Emlen DJ, et al. Juvenile hormone regulates extreme mandible growth in male stag beetles. PLoS One. 2011;**6**:e21139. DOI: 10.1371/journal.pone.0021139

[14] Goodman WG, Cusson M. The juvenile hormones. In: Insect Endocrinology. Academic Press; 2012. pp. 310-365. DOI: 10.1016/B978-0-12-384749-2.10008-1

[15] Williams CM. The juvenile hormone. II. Its role in the endocrine control of molting, pupation, and adult development in the cecropia silkworm. Biological Bulletin. 1961;**121**:572-585. DOI: 10.2307/1539456

[16] Riddiford LM. Hormonal control of insect epidermal cell commitment *in vitro*. Nature. 1976;**259**:115-117. DOI: 10.1038/259115a0

[17] Zitnan D, Kim YJ, Zitnanová I, Roller L, Adams ME. Complex steroid-peptide-receptor cascade controls insect ecdysis. General and Comparative Endocrinology. 2007;**153**:88-96. DOI: 10.1016/j.ygcen.2007.04.002

[18] Qu Z, Kenny NJ, Lam HM, Chan TF, Chu KH, Bendena WG, et al. How did arthropod sesquiterpenoids and ecdysteroids arise? Comparison of hormonal pathway genes in noninsect arthropod genomes. Genome Biology and Evolution. 2015;7:1951-1959. DOI: 10.1093/gbe/evv120

[19] Smith SL, Bollenbacher WE, Gilbert LI. Ecdysone 20-monooxygenase activity during larval-pupal development of *Manduca sexta*. Molecular and Cellular Endocrinology. 1983;**31**:227-251. DOI: 10.1016/0303-7207(83)90151-x

[20] Daimon T, Shinoda T. Function, diversity, and application of insect juvenile hormone epoxydases (CYP15). Biotechnology and Applied Biochemistry. 2013;**60**:82-91. DOI: 10.1002/bab.1058

[21] Sin YW, Kenny NJ, Qu Z, Chan KW, Chan KW, Cheong SP, et al. Identification of putative ecdysteroid and juvenile hormone pathway genes in the shrimp *Neocaridina denticulata*. General and Comparative Endocrinology. 2015;**214**:167-176. DOI: 10.1016/j.ygcen.2014.07.018

[22] Toyota K, Miyakawa H, Hiruta C, Furuta K, Ogino Y, Shinoda T, et al. Methyl farnesoate synthesis is necessary for the environmental sex determination in the water flea *Daphnia pulex*. Journal of Insect Physiology. 2015;**80**:22-30. DOI: 10.1016/j.jinsphys.2015.02.002

[23] Laufer H, Biggers WJ. Unifying concepts learned from methyl farnesoate for invertebrate reproduction and post-embryonic development. American Zoologist. 2001;**41**:442-457. DOI: 10.1093/icb/41.3.442

[24] McWilliam PS, Phillips BF. Spiny lobster development: Mechanisms inducing metamorphosis to the puerulus: A review. Reviews in Fish Biology and Fisheries. 2007;**17**:615-632. DOI: 10.1007/s11160-007-9067-5

[25] Katayama H, Ohira T, Nagasawa H. Crustacean peptide hormones: Structure, gene expression and function. Aqua-BioScience Monographs. 2013;**6**:49-90. DOI: 10.5047/absm.2013.00602.0049

[26] Møller OS, Anger K, Guerao G. Patterns of larval development. In: Developmental Biology and Larval Ecology: The Natural History of the Crustacea, Volume 7. 2020. DOI: 10.1093/oso/9780190648954.003.0006. Available from: https://oxford.universitypressscholarship.com/view/10.1093/oso/9780190648954.001.0001/oso-9780190648954-chapter-6

[27] Toyota K, Yamane F, Ohira T. Impacts of methyl farnesoate and 20-hydroxyecdysone on larval mortality and metamorphosis in the kuruma prawn *Marsupenaeus japonicus*. Frontiers in Endocrinology. 2020;**11**:475. DOI: 10.3389/fendo.2020.00475

[28] Abdu U, Takac P, Laufer H, Sagi A. Effect of methyl farnesoate on late larval development and metamorphosis in the prawn *Macrobrachium rosenbergii* (Decapoda, Palaemonidae): A juvenoid-like effect? Biology Bulletin. 1998;**195**:112-119. DOI: 10.2307/1542818

[29] McKenney CL, Celestial DM. Variations in larval growth and metabolism of an estuarine shrimp *Palaemonetes pugio* during toxicosis by an insect growth regulator. Comparative Biochemistry and Physiology Part C: Comparative Pharmacology. 1993;**105**:239-245. DOI: 10.1016/0742-8413(93)90201-U

[30] Smith PA, Clare AS, Rees HH, Prescott MC, Wainwright G, Thorndyke MC. Identification of methyl farnesoate in the cypris larva of the barnacle, *Balanus amphitrite*, and its role as a juvenile hormone. Insect Biochemistry and Molecular Biology. 2000;**30**:885-890. DOI: 10.1016/s0965-1748(00)00062-x

[31] Yamamoto H, Okino T, Yoshimura E, Tachibana A, Shimizu K, Fusetani N. Methyl farnesoate induces larval metamorphosis of the barnacle, *Balanus amphitrite* via protein kinase C activation. Journal of Experimental Zoology. 1997;**278**:349-355. DOI: 10.1002/(sici)1097-010x(19970815)278:63.0.co;2-o

[32] Freeman JA, Costlow JD. The molt cycle and its hormonal control in *Rhithropanopeus harrisii* larvae. Developmental Biology. 1980;**74**:479-485. DOI: 10.1016/0012-1606(80)90447-9

[33] Dan S, Kaneko T, Takeshima S, Ashidate M, Hamasaki K. Eyestalk ablation affects larval morphogenesis in the swimming crab *Portunus trituberculatus* during metamorphosis into megalopae. Sexuality and Early Development in Aquatic Organisms. 2014;**1**:57-73. DOI: 10.3354/sedao00007

[34] Synder MJ, Chang ES. Effects of eyestalk ablation on larval molting rates and morphological development of the american lobster *homarus americanus*. Biological Bulletin. 1986;**170**:232-243. DOI: 10.2307/1541805

[35] Laufer H, Takac P, Ahl JSB, Laufer MR. Methyl farnesoate and the effect of eyestalk ablation on the morphogenesis of the juvenile female spider crab *Libinia emarginata*. Invertebrate Reproduction and Development. 1997;**31**:63-68. DOI: 10.1080/07924259.1997.9672564

[36] Ventura T, Fitzgibbon QP, Battaglene SC, Elizur A. Redefining metamorphosis in spiny lobsters: Molecular analysis of the phyllosoma to puerulus transition in *Sagmariasus verreauxi*. Scientific Reports. 2015;**5**:13537. DOI: 10.1038/srep13537

[37] Farley RD. Developmental changes in the embryo, pronymph, and first molt of the scorpion *Centruroides vittatus* (Scorpiones: Buthidae). Journal of Morphology. 2005;**265**:1-27. DOI: 10.1002/jmor.10227

[38] Lourenço WR. Scorpions and life-history strategies: From evolutionary dynamics toward the scorpionism problem. Journal of Venomous Animals and Toxins Including Tropical Diseases. 2018;**24**:19. DOI: 10.1186/s40409-018-0160-0

[39] Polis GA, Farley RD. Characteristics and environmental determinants of natality, growth and maturity in a natural population of the desert scorpion, *Paruroctonus mesaensis* (Scorpionida: Vaejovidae). Journal of Zoology.

1979;**187**:517-542. DOI: 10.1111/J.1469-7998.1979.TB03385.X

[40] Polis GA, Sissom WD. Life history. In: Polis GA, editor. The Biology of Scorpions. Stanford: Stanford Univ. Press; 1990. pp. 161, 587-223

[41] Landulfo GA, Pevidor LV, Luz HR, Faccini JLH, Nunes PH, Barros-Battesti DM. Description of nymphal instars of *Ornithodoros mimon* Kohls, Clifford & Jones, 1969 (Acari: Argasidae). Zootaxa. 2013;**3710**:12. DOI: 10.11646/zootaxa.3710.2.4

[42] Santos VT, Ribeiro L, Fraga A, de Barros CM, Campos E, Moraes J, et al. The embryogenesis of the tick *Rhipicephalus (Boophilus) microplus*: The establishment of a new chelicerate model system. Genesis. 2013;**51**:803-818. DOI: 10.1002/dvg.22717

[43] Quade FSC, Holtzheimer J, Frohn J, Töpperwien M, Salditt T, Prpic NM. Formation and development of the male copulatory organ in the spider *Parasteatoda tepidariorum* involves a metamorphosis-like process. Scientific Reports. 2019;**9**:6945. DOI: 10.1038/s41598-019-43192-9

[44] Lafont R, Mathieu M. Steroids in aquatic invertebrates. Ecotoxicology. 2007;**16**:109-130. DOI: 10.1007/s10646-006-0113-1

[45] Jindra M, Palli SR, Riddiford LM. The juvenile hormone signaling pathway in insect development. Annual Review of Entomology. 2013;**58**:181-204. DOI: 10.1146/annurev-ento-120811-153700

[46] Miyakawa H, Toyota K, Sumiya E, Iguchi T. Comparison of JH signaling in insects and crustaceans. Current Opinion in Insect Science. 2014;**1**:81-87. DOI: 10.1016/j.cois.2014.04.006

[47] Bonaric JC, De Reggi M. Changes in ecdysone levels in the spider *Pisaura mirabilis nymphs* (Araneae, Pisauridae). Experimentia. 1977;**33**:1664-1665. DOI: 10.1007/BF01934060

[48] Delbecque JP, Diehl PA, O'Connor JD. Presence of ecdysone and ecdysterone in the tick *Amblyomma hebraeum* Koch. Experientia. 1978;**34**: 1379-1381

[49] Rees HH. Hormonal control of tick development and reproduction. Parasitology. 2004;**129**:127-143. DOI: 10.1017/s003118200400530x

[50] Sawadro M, Bednarek A, Babczyńska A. The current state of knowledge on the neuroactive compounds that affect the development, mating and reproduction of spiders (Araneae) compared to insects. Invertebrate Neuroscience. 2017;**17**:4. DOI: 10.1007/s10158-017-0197-8

[51] Winget RR, Herman WS. Occurrence of ecdysone in the blood of the chelicerate arthropod, *Limulus polyphemus*. Experientia. 1976;**32**:1345-1346. DOI: 10.1007/BF01953131

[52] Zhu XX, Oliver JH Jr, Dotson EM. Epidermis as the source of ecdysone in an argasid tick. Proceedings of the National Academy of Sciences of the United States of America. 1991;**88**:3744-3747. DOI: 10.1073/pnas.88.9.3744

[53] Guo X, Harmon MA, Laudet V, Mangelsdorf DJ, Palmer MJ. Isolation of a functional ecdysteroid receptor homologue from the ixodid tick *Amblyomma americanum* (L.). Insect Biochemistry and Molecular Biology. 1997;**27**:945-962. DOI: 10.1016/s0965-1748(97)00075-1

[54] Guo X, Xu Q, Harmon MA, Jin X, Laudet V, Mangelsdorf DJ, et al.

Isolation of two functional retinoid X receptor subtypes from the Ixodid tick, *Amblyomma americanum* (L.). Molecular and Cellular Endocrinology. 1998;**139**:45-60. DOI: 10.1016/S0303-7207(98)00073-2

[55] Honda Y, Ishiguro W, Ogihara MH, Kataoka H, Taylor D. Identification and expression of nuclear receptor genes and ecdysteroid titers during nymphal development in the spider *Agelena sylvatica*. General and Comparative Endocrinology. 2017;**247**:183-198. DOI: 10.1016/j.ygcen.2017.01.032

[56] Horigane M, Ogihara K, Nakajima Y, Shinoda T, Taylor D. Cloning and expression of the ecdysteroid receptor during ecdysis and reproduction in females of the soft tick, *Ornithodoros moubata* (Acari: Argasidae). Insect Molecular Biology. 2007;**16**:601-612. DOI: 10.1111/j.1365-2583.2007.00754.x

[57] Horigane M, Ogihara K, Nakajima Y, Taylor D. Isolation and expression of the retinoid X receptor from last instar nymphs and adult females of the soft tick *Ornithodoros moubata* (Acari: Argasidae). General and Comparative Endocrinology. 2008;**156**:298-311. DOI: 10.1016/j.ygcen.2008.01.021

[58] Miyashita M, Matsushita K, Nakamura S, Akahane S, Nakagawa Y, Miyagawa H. LC/MS/MS identification of 20-hydroxyecdysone in a scorpion (*Liocheles australasiae*) and its binding affinity to *in vitro*-translated molting hormone receptors. Insect Biochemistry and Molecular Biology. 2011;**41**:932-937. DOI: 10.1016/j.ibmb.2011.09.002

[59] Nakagawa Y, Sakai A, Magata F, Ogura T, Miyashita M, Miyagawa H. Molecular cloning of the ecdysone receptor and the retinoid X receptor from the scorpion *Liocheles australasiae*. The FEBS Journal. 2007;**274**:6191-6203. DOI: 10.1111/j.1742-4658.2007.06139.x

[60] Bonaric JC. Effects of ecdysterone on the molting mechanisms and duration of the intermolt period in *Pisaura mirabilis* Cl. General and Comparative Endocrinology. 1976;**30**:267-272. DOI: 10.1016/0016-6480(76)90077-0

[61] Jegla TC, Costlow JD, Alspaugh J. Effects of ecdysones and some synthetic analogs on horseshoe crab larvae. General and Comparative Endocrinology. 1972;**19**:159-166. DOI: 10.1016/0016-6480(72)90016-0

[62] Krishnakumaran A, Schneiderman HA. Chemical control of moulting in arthropods. Nature. 1968;**220**:601-603. DOI: 10.1038/220601a0

[63] Krishnakumaran A, Schneiderman HA. Control of molting in mandibulate and chelicerate arthropods by ecdysones. Nature. 1970;**139**:520-538. DOI: 10.2307/1540371

[64] Nicewicz AW, Sawadro MK, Nicewicz Ł, Babczyńska AI. Juvenile hormone in spiders. Is this the solution to a mystery? General and Comparative Endocrinology. 2021;**308**:113781. DOI: 10.1016/j.ygcen.2021.113781

[65] Zhu J, Khalil SM, Mitchell RD, Bissinger BW, Egekwu N, Sonenshine DE, et al. Mevalonate-farnesal biosynthesis in ticks: Comparative synganglion transcriptomics and a new perspective. PLoS One. 2016;**11**:e0141084. DOI: 10.1371/journal.pone.0141084

[66] Verhulst EC, van de Zande L. Double nexus--Doublesex is the connecting element in sex determination. Briefings Functional Genomics. 2015;**14**:396-406. DOI: 10.1093/bfgp/elv005

[67] Nagahama Y, Chakraborty T, Paul-Prasanth B, Ohta K, Nakamura M. Sex determination, gonadal sex differentiation, and plasticity in vertebrate species. Physiological Reviews. 2021;**101**:1237-1308. DOI: 10.1152/physrev.00044.2019

[68] Verhulst EC, van de Zande L, Beukeboom LW. Insect sex determination: It all evolves around transformer. Current Opinion in Genetics and Development. 2010;**20**:376-383. DOI: 10.1016/j.gde.2010.05.001

[69] Wilhelm D, Palmer S, Koopman P. Sex determination and gonadal development in mammals. Physiological Reviews. 2007;**87**:1-28. DOI: 10.1152/physrev.00009.2006

[70] Kopp A. Dmrt genes in the development and evolution of sexual dimorphism. Trends in Genetics. 2012;**28**:175-184. DOI: 10.1016/j.tig.2012.02.002

[71] Matson C, Zarkower D. Sex and the singular DM domain: Insights into sexual regulation, evolution and plasticity. Nature Review Genetics. 2012;**13**:163-174. DOI: 10.1038/nrg3161

[72] Picard MA, Cosseau C, Mouahid G, Duval D, Grunau C, Toulza È, et al. The roles of Dmrt (double sex/male-abnormal-3 related transcription factor) genes in sex determination and differentiation mechanisms: Ubiquity and diversity across the animal kingdom. Comptes Rendus Biologies. 2015;**338**:451-462. DOI: 10.1016/j.crvi.2015.04.010

[73] Volff JN, Zarkower D, Bardwell VJ, Schartl M. Evolutionary dynamics of the DM domain gene family in metazoans. Journal of Molecular Evolution. 2003;**57**:241-249. DOI: 10.1007/s00239-003-0033-0

[74] Pane A, Salvemini M, Bovi PD, Polito C, Saccone G. The *transformer* gene in *Ceratitis capitata* provides a genetic basis for selecting and remembering the sexual fate. Development. 2002;**129**:3715-3725. DOI: 10.1242/dev.129.15.3715

[75] Prakash A, Monteiro A. Molecular mechanisms of secondary sexual trait development in insects. Current Opinion in Insect Science. 2016;**17**:40-48. DOI: 10.1016/j.cois.2016.06.003

[76] Kato Y, Kobayashi K, Watanabe H, Iguchi T. Environmental sex determination in the branchiopod crustacean *Daphnia magna*: Deep conservation of a *doublesex* gene in the sex-determining pathway. PLoS Genetics. 2011;7:e1001345. DOI: 10.1371/journal.pgen.1001345

[77] Nong QD, Matsuura T, Kato Y, Watanabe H. Two *Doublesex1* mutants revealed a tunable gene network underlying intersexuality in *Daphnia magna*. PLoS One. 2020;**15**:e0238256. DOI: 10.1371/journal.pone.0238256

[78] Toyota K, Kato Y, Sato M, Sugiura N, Miyagawa S, Miyakawa H, et al. Molecular cloning of doublesex genes of four cladocera (water flea) species. BMC Genomics. 2013;**14**:239. DOI: 10.1186/1471-2164-14-239

[79] Toyota K, Miyakawa H, Hiruta C, Sato T, Katayama H, Ohira T, et al. Sex determination and differentiation in decapod and cladoceran crustaceans: An overview of endocrine regulation. Genes. 2021;**12**:305. DOI: 10.3390/genes12020305

[80] Charniaux-Cotton H. Discovery in an amphipod crustacean (*Orchestia gammarella*) of an endocrine gland responsible for the differentiation of primary and secondary male sex characteristics. Comptes rendus

de l'Académie des Sciences. Paris. 1954;**239**:780-782

[81] Katakura Y. Transformation of ovary into testis following implantation of androgenic glands in *Armadillidium vulgare*, an isopod crustacean. Annotationes zoologicae Japonenses. 1960;**33**:241-244

[82] Suzuki S. Androgenic gland hormone is a sex-reversing factor but cannot be a sex-determining factor in the female crustacean isopods *Armadillidium vulgare*. General and Comparative Endocrinology. 1999;**115**:370-378. DOI: 10.1006/gcen.1999.7324

[83] Katakura Y, Hasegawa Y. Masculinization of females of the isopod crustacean, *Armadillidium vulgare*, following injections of an active extract of the androgenic gland. General and Comparative Endocrinology. 1983;**49**:57-62. DOI: 10.1016/0016-6480(83)90007-2

[84] Hasegawa Y, Haino-Fukushima K, Katakura Y. Isolation and properties of androgenic gland hormone from the terrestrial isopod, *Armadillidium vulgare*. General and Comparative Endocrinology. 1987;**67**:101-110. DOI: 10.1016/0016-6480(87)90209-7

[85] Okuno A, Hasegawa Y, Nagasawa H. Purification and properties of androgenic gland hormone from the terrestrial isopod *Armadillidium vulgare*. Zoological Science. 1997;**14**:837-842. DOI: 10.2108/zsj.14.837

[86] Kato M, Hiruta C, Tochinai S. Androgenic gland implantation induces partial masculinization in marmorkrebs *Procambarus fallax* f. *virginalis*. Zoological Science. 2015;**32**:459-464. DOI: 10.2108/zs150028

[87] Katayama H, Toyota K, Tanaka H, Ohira T. Chemical synthesis and functional evaluation of the crayfish insulin-like androgenic gland factor. Bioorganic Chemistry. 2022;**122**:105738. DOI: 10.1016/j.bioorg.2022.105738

[88] Khalaila I, Manor R, Weil S, Granot Y, Keller R, Sagi A. The eyestalk-androgenic gland-testis endocrine axis in the crayfish *Cherax quadricarinatus*. General and Comparative Endocrinology. 2002;**127**:147-156. DOI: 10.1016/s0016-6480(02)00031-x

[89] Sroyraya M, Chotwiwatthanakun C, Stewart MJ, Soonklang N, Kornthong N, Phoungpetchara I, et al. Bilateral eyestalk ablation of the blue swimmer crab, *Portunus pelagicus*, produces hypertrophy of the androgenic gland and an increase of cells producing insulin-like androgenic gland hormone. Tissue and Cell. 2010;**42**:293-300. DOI: 10.1016/j.tice.2010.07.003

[90] Aizen J, Chandler JC, Fitzgibbon QP, Sagi A, Battaglene SC, Elizur A, et al. Production of recombinant insulin-like androgenic gland hormones from three decapod species: *In vitro* testicular phosphorylation and activation of a newly identified tyrosine kinase receptor from the eastern spiny lobster, *Sagmariasus verreauxi*. General and Comparative Endocrinology. 2016;**229**:8-18. DOI: 10.1016/j.ygcen.2016.02.013

[91] Chandler JC, Aizen J, Elizur A, Hollander-Cohen L, Battaglene SC, Ventura T. Discovery of a novel insulin-like peptide and insulin binding proteins in the eastern rock lobster *Sagmariasus verreauxi*. General and Comparative Endocrinology. 2015;**215**:76-87. DOI: 10.1016/j.ygcen.2014.08.018

[92] Rosen O, Weil S, Manor R, Roth Z, Khalaila I, Sagi A. A crayfish insulin-like-binding protein: Another piece in the androgenic gland insulin-like hormone puzzle is revealed. Journal of Biological

Chemistry. 2013;**288**:22289-22298. DOI: 10.1074/jbc.M113.484279

[93] Li S, Li F, Yu K, Xiang J. Identification and characterization of a doublesex gene which regulates the expression of insulin-like androgenic gland hormone in *Fenneropenaeus chinensis*. Gene. 2018;**649**:1-7. DOI: 10.1016/j.gene.2018.01.043

[94] Zheng J, Cai L, Jia Y, Chi M, Cheng S, Liu S, et al. Identification and functional analysis of the *doublesex* gene in the redclaw crayfish, *Cherax quadricarinatus*. Gene Expression Patterns. 2020;**37**:119129. DOI: 10.1016/j.gep.2020.119129

[95] Zmora N, Chung JS. A novel hormone is required for the development of reproductive phenotypes in adult female crabs. Endocrinology. 2014;**155**:230-239. DOI: 10.1210/en.2013-1603

[96] Veenstra JA. The power of next-generation sequencing as illustrated by the neuropeptidome of the crayfish *Procambarus clarkii*. General and Comparative Endocrinology. 2015;**224**:84-95. DOI: 10.1016/j.ygcen.2015.06.013

[97] Veenstra JA. Similarities between decapod and insect neuropeptidomes. PeerJ. 2016;**4**:e2043. DOI: 10.7717/peerj.2043

[98] Oliphant A, Alexander JL, Swain MT, Webster SG, Wilcockson DC. Transcriptomic analysis of crustacean neuropeptide signaling during the moult cycle in the green shore crab, *Carcinus maenas*. BMC Genomics. 2018;**19**:711. DOI: 10.1186/s12864-018-5057-3

[99] Liu A, Liu J, Liu F, Huang Y, Wang G, Ye H. Crustacean female sex hormone from the mud crab *Scylla paramamosain* is highly expressed in prepubertal males and inhibits the development of androgenic gland. Frontiers in Physiology. 2018;**9**:924. DOI: 10.3389/fphys.2018.00924

[100] Jiang Q, Lu B, Lin D, Huang H, Chen X, Ye H. Role of crustacean female sex hormone (CFSH) in sex differentiation in early juvenile mud crabs, *Scylla paramamosain*. General and Comparative Endocrinology. 2020;**289**:113383. DOI: 10.1016/j.ygcen.2019.113383

[101] Jiang Q, Lu B, Wang G, Ye H. Transcriptional inhibition of Sp-IAG by crustacean female sex hormone in the mud crab, *Scylla paramamosain*. International Journal of Molecular Sciences. 2020;**21**:5300. DOI: 10.3390/ijms21155300

[102] Kotaka S, Ohira T. cDNA cloning and *in situ* localization of a crustacean female sex hormone-like molecule in the kuruma prawn *Marsupenaeus japonicus*. Fisheries Science. 2018;**84**:53-60. DOI: 10.1007/s12562-017-1152-7

[103] Powell D, Knibb W, Remilton C, Elizur A. *De-novo* transcriptome analysis of the banana shrimp (*Fenneropenaeus merguiensis*) and identification of genes associated with reproduction and development. Marine Genomics. 2015;**22**:71-78. DOI: 10.1016/j.margen.2015.04.006

[104] Toullec JY, Corre E, Mandon P, Gonzalez-Aravena M, Ollivaux C, Lee CY. Characterization of the neuropeptidome of a Southern Ocean decapod, the Antarctic shrimp *Chorismus antarcticus*: Focusing on a new decapod ITP-like peptide belonging to the CHH peptide family. General and Comparative Endocrinology. 2017;**252**:60-78. DOI: 10.1016/j.ygcen.2017.07.015

[105] Ventura T, Cummins SF, Fitzgibbon Q, Battaglene S, Elizur A. Analysis of the central nervous system transcriptome of the eastern rock lobster *Sagmariasus verreauxi* reveals its putative neuropeptidome. PLoS One. 2014;**9**:e97323. DOI: 10.1371/journal.pone.0097323

[106] Suwansa-Ard S, Thongbuakaew T, Wang T, Zhao M, Elizur A, Hanna PJ, et al. *In silico* neuropeptidome of female *Macrobrachium rosenbergii* based on transcriptome and peptide mining of eyestalk, central nervous system and ovary. PLoS One. 2015;**10**:e0123848. DOI: 10.1371/journal.pone.0123848

[107] Thongbuakaew T, Sumpownon C, Engsusophon A, Kornthong N, Chotwiwatthanakun C, Meeratana P, et al. Characterization of prostanoid pathway and the control of its activity by the eyestalk optic ganglion in the female giant freshwater prawn, *Macrobrachium rosenbergii*. Heliyon. 2021;**7**:e05898. DOI: 10.1016/j.heliyon.2021.e05898

[108] Liu F, Shi W, Huang L, Wang G, Zhu Z, Ye H. Roles of crustacean female sex hormone 1a in a protandric simultaneous hermaphrodite shrimp. Fronters in Marine Science. 2021;**8**:791965. DOI: 10.3389/fmars.2021.791965

[109] Liu F, Shi W, Ye H, Zeng C, Zhu Z. Insulin-like androgenic gland hormone 1 (IAG1) regulates sexual differentiation in a hermaphrodite shrimp through feedback to neuroendocrine factors. General and Comparative Endocrinology. 2021;**303**:113706. DOI: 10.1016/j.ygcen.2020.113706

[110] Nguyen TV, Cummins SF, Elizur A, Ventura T. Transcriptomic characterization and curation of candidate neuropeptides regulating reproduction in the eyestalk ganglia of the Australian crayfish, *Cherax quadricarinatus*. Scientific Reports. 2016;**6**:38658. DOI: 10.1038/srep38658

[111] Tsutsui N, Kotaka S, Ohira T, Sakamoto T. Characterization of distinct ovarian isoform of crustacean female sex hormone in the kuruma prawn *Marsupenaeus japonicus*. Comparative Biochemistry and Physiology - Part A: Molecular and Integrative Physiology. 2018;**217**:7-16. DOI: 10.1016/j.cbpa.2017.12.009

[112] Kuntner M, Coddington JA. Sexual size dimorphism: Evolution and perils of extreme phenotypes in spiders. Annual Review of Entomology. 2020;**65**:57-80. DOI: 10.1146/annurev-ento-011019-025032

[113] Šet J, Turk E, Golobinek R, Lokovšek T, Gregorič M, Lebrón SGQ, et al. Sex-specific developmental trajectories in an extremely sexually size dimorphic spider. Die Naturwissenschaften. 2021;**108**:54. DOI: 10.1007/s00114-021-01754-w

[114] Cordellier M, Schneider JM, Uhl G, Posnien N. Sex differences in spiders: From phenotype to genomics. Development Genes and Evolution. 2020;**230**:155-172. DOI: 10.1007/s00427-020-00657-6

[115] Girard MB, Kasumovic MM, Elias DO. Multi-modal courtship in the peacock spider, *Maratus volans* (O.P.-Cambridge, 1874). PLoS One. 2011;**6**:e25390. DOI: 10.1371/journal.pone.0025390

[116] Pomerantz AF, Hoy MA, Kawahara AY. Molecular characterization and evolutionary insights into potential sex-determination genes in the western orchard predatory mite

Metaseiulus occidentalis (Chelicerata: Arachnida: Acari: Phytoseiidae). Journal of Biomolecular Structure and Dynamics. 2015;**33**:1239-1253. DOI: 10.1080/07391102.2014.941402

[117] Pomerantz AF, Hoy MA. Expression analysis of *drosophila doublesex*, *transformer-2*, *intersex*, *fruitless-like*, and *vitellogenin* homologs in the parahaploid predator *Metaseiulus occidentalis* (Chelicerata: Acari: Phytoseiidae). Experimental and Applied Acarology. 2015;**65**:1-16. DOI: 10.1007/s10493-014-9855-2

[118] Schwager EE, Sharma PP, Clarke T, Leite DJ, Wierschin T, Pechmann M, et al. The house spider genome reveals an ancient whole-genome duplication during arachnid evolution. BMC Biology. 2017;**15**:62. DOI: 10.1186/s12915-017-0399-x

[119] Gruzin M, Mekheal M, Ruhlman K, Winkowski M, Petko J. Developmental expression of doublesex-related transcripts in the common house spider, *Parasteatoda tepidariorum*. Gene Expression Patterns. 2020;**35**:119101. DOI: 10.1016/j.gep.2020.119101

[120] Melzer RR, Spitzner F, Šargač Z, Hörnig MK, Krieger J, Haug C, et al. Methods to study organogenesis in decapod crustacean larvae II: Analysing cells and tissues. Helgoland Marine Research. 2021;**75**:2. DOI: 10.1186/s10152-021-00547-y

[121] Torres G, Melzer RR, Spitzner F, Šargač Z, Harzsch S, Gimenez L. Methods to study organogenesis in decapod crustacean larvae. I. Larval rearing, preparation, and fixation. Helgoland Marine Research. 2021;**75**:3. DOI: 10.1186/s10152-021-00548-x

[122] Bruce HS, Patel NH. Knockout of crustacean leg patterning genes suggests that insect wings and body walls evolved from ancient leg segments. Nature Ecology & Evolution. 2020;**4**:1703-1712

[123] Clark-Hachtel CM, Tomoyasu Y. Two sets of candidate crustacean wing homologues and their implication for the origin of insect wings. Nature Ecology & Evolution. 2020;**4**:1694-1702. DOI: 10.1038/s41559-020-1257-8

[124] Sun DA, Patel NH. The amphipod crustacean *Parhyale hawaiensis*: An emerging comparative model of arthropod development, evolution, and regeneration. Wiley Interdisciplinary Reviews: Developmental Biology. 2019;**8**:e355. DOI: 10.1002/wdev.355

[125] Toyota K, Miyagawa S, Ogino Y, Iguchi T: Microinjection-based RNA interference method in the water flea, *Daphnia pulex* and *Daphnia magna*. In: RNA Interference; InTech: London; 2016.p. 119-135

[126] Guo L, Xie W, Liu Y, Yang X, Xia J, Wang S, et al. Identification and characterization of *doublesex* in *Bemisia tabaci*. Insect Molecular Biology. 2018;**27**:620-632. DOI: doi.org/10.1111/imb.12494

[127] Miyazaki S, Fujiwara K, Kai K, Masuoka Y, Gotoh H, Niimi T, et al. Evolutionary transition of *doublesex* regulation from sex-specific splicing to male-specific transcription in termites. Scientific Reports. 2021;**11**:15992. DOI: 10.1038/s41598-021-95423-7

[128] Wexler J, Delaney EK, Belles X, Schal C, Wada-Katsumata A, Amicucci MJ, et al. Hemimetabolous insects elucidate the origin of sexual development via alternative splicing. eLife. 2019;**8**:e47490. DOI: 10.7554/eLife.47490

[129] Chipman AD, Ferrier DEK, Brena C, Qu J, Hughes DST, Schröder R, et al. The

first Myriapod genome sequence reveals conservative arthropod gene content and genome organisation in the centipede *Strigamia maritima*. PLoS Biology. 2014;**12**:e1002005. DOI: 10.1371/journal.pbio.1002005

[130] Schumann I, Kenny N, Hui J, Hering L, Mayer G. Halloween genes in panarthropods and the evolution of the early moulting pathway in Ecdysozoa. Royal Society Open Science. 2018;**5**:180888. DOI: 10.1098/rsos.180888

[131] Kiseleva AP, Kiselev GO, Nikolaeva VO, Seisenbaeva G, Kessler V, Krivoshapkin PV, et al. Hybrid spider silk with inorganic aanomaterials. Nanomaterials. 2020;**10**:1853. DOI: 10.3390/nano10091853

[132] Kluge JA, Rabotyagova O, Leisk GG, Kaplan DL. Spider silks and their applications. Trends in Biotechnology. 2008;**26**:244-251. DOI: 10.1016/j.tibtech.2008.02.006

[133] Rising A, Johansson J. Toward spinning artificial spider silk. Nature Chemical Biology. 2015;**11**:309-315. DOI: 10.1038/nchembio.1789

[134] Heim M, Keerl D, Scheibel T. Spider silk: From soluble protein to extraordinary fiber. Angewandte Chemie. 2009;**48**:3584-3596. DOI: 10.1002/anie.200803341

[135] Kiseleva AP, Krivoshapkin PV, Krivoshapkina EF. Recent advances in development of functional spider silk-based hybrid materials. Frontiers in Chemistry. 2020;**8**:554. DOI: 10.3389/fchem.2020.00554

[136] Roberts AD, Finnigan W, Kelly PP, Faulkner M, Breitling R, Takano E, et al. Non-covalent protein-based adhesives for transparent substrates-bovine serum albumin vs. recombinant spider silk. Materials Today Biology. 2020;**7**:100068. DOI: 10.1016/j.mtbio.2020.100068

[137] Bakhshandeh B, Nateghi SS, Gazani MM, Dehghani Z, Mohammadzadeh F. A review on advances in the applications of spider silk in biomedical issues. International Journal of Biological Macromolecules. 2021;**192**:258-271. DOI: 10.1016/j.ijbiomac.2021.09.201

[138] Spiess K, Lammel A, Scheibel T. Recombinant spider silk proteins for applications in biomaterials. Macromolecular Bioscience. 2010;**10**:998-1007. DOI: 10.1002/mabi.201000071

[139] Radtke C. Natural occurring silks and their analogues as materials for nerve conduits. International Journal of Molecular Sciences. 2016;**17**:1754. DOI: 10.3390/ijms17101754

[140] Salehi S, Koeck K, Scheibel T. Spider silk for tissue engineering applications. Molecules. 2020;**25**:737. DOI: 10.3390/molecules25030737

[141] Abdel-Rahman MA, Harrison PL, Strong PN. Snapshots of scorpion venomics. Journal of Arid Environments. 2015;**112**:170-176. DOI: 10.1016/j.jaridenv.2014.01.007

[142] Ortiz E, Gurrola GB, Schwartz EF, Possani LD. Scorpion venom components as potential candidates for drug development. Toxicon. 2015;**93**:125-135. DOI: 10.1016/j.toxicon.2014.11.233

[143] Saez NJ, Herzig V. Versatile spider venom peptides and their medical and agricultural applications. Toxicon. 2019;**158**:109-126. DOI: 10.1016/j.toxicon.2018.11.298

[144] Cohen G, Burks SR, Frank JA. Chlorotoxin—A multimodal imaging

platform for targeting glioma tumors. Toxins. 2018;**10**:496. DOI: 10.3390/toxins10120496

[145] De Sousa L, Borges A, Vásquez-Suárez A, den Camp HJMO, Chadee-Burgos RI, Romero-Bellorín M, et al. Differences in venom toxicity and antigenicity between females and males *Tityus nororientalis* (Buthidae) scorpions. Journal of Venom Research. 2010;**1**:61-70

[146] Gao S, Wu F, Chen X, Yang Y, Zhu Y, Xiao L, et al. Sex-biased gene expression of *Mesobuthus martensii* collected from Gansu Province, China, reveals their different therapeutic potentials. Evidence-Based Complementary and Alternative Medicine. 2021;**2021**:1967158. DOI: 10.1155/2021/1967158

[147] de Oliveira KC. Gonçalves de Andrade RM, piazza RM, Ferreira JM Jr, van den berg CW, Tambourgi DV: Variations in *Loxosceles* spider venom composition and toxicity contribute to the severity of envenomation. Toxicon. 2005;**45**:421-429. DOI: 10.1016/j.toxicon.2004.08.022

[148] Uribe CJI, Jiménez Vargas JM, Ferreira Batista CV, Zamudio Zuñiga F, Possani LD. Comparative proteomic analysis of female and male venoms from the Mexican scorpion *Centruroides limpidus*: Novel components found. Toxicon. 2017;**125**:91-98. DOI: 10.1016/j.toxicon.2016.11.256

[149] Yamaji N, Dai L, Sugase K, Andriantsiferana M, Nakajima T, Iwashita T. Solution structure of IsTX. A male scorpion toxin from *Opisthacanthus madagascariensis* (Ischnuridae). European Journal of Biochemistry. 2004;**271**:3855-3864. DOI: 10.1111/j.1432-1033.2004.04322.x

Chapter 2

Spatiotemporal Dynamics of *Bemisia tabaci* MEAM1 (Hemiptera: Aleyrodidae) in Commercial Soybean Crops

Luciana Barboza Silva, Raimundo Henrique Ferreira Rodrigues, Thiago Ferreira Rodrigues, Maria Carolina Farias e Silva, Edivania de Araujo Lima and José Wellington Batista Lopes

Abstract

Spatiotemporal dynamics studies of crop pests enable the determination of the colonization pattern and dispersion of these insects in the landscape. Geostatistics is an efficient tool for these studies: to determine the spatial distribution pattern of the pest in the crops and to make maps that represent this situation. The aim of this study was to determine the spatiotemporal distribution of *B. tabaci* adults and nymphs in commercial soybean crops from planting to harvest using geostatistics. Infestation by adults and nymphs of *B. tabaci* started between 30 and 50 days after the emergence of the plants. The maximum population density of ten adults per plant and two nymphs per leaf was registered between 90 and 101 days after plant emergence. The colonization of soybean plants by *B. tabaci* may be divided into three stages: beginning infestation (at the outermost parts of the crop), whole area colonization, and dispersion colonization (when the whole crop area is infested). The density of adult insects was positively correlated with rainfall and relative humidity. Wind speed positively affected the dispersion of adult whiteflies. The distribution pattern of *B. tabaci* in the soybean crop was aggregated.

Keywords: Aleyrodidae, dispersion, Geostatistics, *Glycine max*, Hemiptera, whitefly

1. Introduction

Bemisia tabaci (Gennadius) Middle East -Asia Minor 1 - MEAM1 (Hemiptera: Aleyrodidae), commonly known as whitefly, is one of the most invasive and destructive pests for global agriculture, causing significant yield losses in commercial crops [1, 2]. This insect has high fecundity and fertility rates. It is polyphagous and occurs in areas of temperate, subtropical, and tropical climates [3–5]. The whitefly has a wide host range; thus, crops such as cotton, beans, squash, melon, and tomato may serve as alternate host plants for this species infesting soybean crops.

 IntechOpen

The whitefly causes several problems in Brazilian crops, and the insect may cause direct damage by sucking phloem sap. Indirect damage is linked to the excretion of honeydew, which serves as a substrate for the growth of opportunistic fungi (Capnodium sp.) [6–8]. *B. tabaci* may inhibit the gathering of carotenoids and chlorophyll affecting the photosynthetic rate of plants [9]. In soybean [*G. max* (L.) Merr., Fabaceae], *B. tabaci* is the vector for the cowpea mild mottle virus (CpMMV) [10].

In recent years, soybean has been highly attacked by high whitefly populations, mainly in the reproductive stages (R1 stage) Until very recently, this species was only seen as an occasional pest, however it is now considered a key pest in soybean. Although the importance of the whitefly *B. tabaci* MEAM1 has increased its incidence in soybean cultivars [11–13], there is no sampling method currently defined for this pest in this crop. The recommendations do not specify the sample numbers, size of area to be sampled, location of the different stages, or the most suitable forms of evaluation. In general, management is based on information obtained from other countries and what has been observed in other crops, such as cotton [14–16] and melon [17], in the state of Arizona, USA. In Brazil [18].

Understanding the spatiotemporal dynamics of pest insects in crops provides important information that can be incorporated into integrated pest management programs [19–22]. The study of the spatiotemporal dynamics of insects using geostatistics can provide important information about the pattern of colonization, aggregation, and dispersion of pests in the field [20, 22–25]. This statistical tool uses a method that characterizes spatial variation by comparing similarities between distant and proximal points.

This technique provides results that allow colonization maps to be produced, zoning the different densities and determining the pattern of spatial distribution of the insects in the field [20, 25]. A sequence of these maps during crop development may indicate areas that demand greater attention for the observation of pest infestation over time. In these spatiotemporal dynamic's studies, plants in the landscape that can serve as pest sources can be identified and located. This may be useful for farmers when planning the location of crops in the landscape or even for predicting where pest attacks will start within the crop area [24, 26, 27].

Considering the importance of studying the spatiotemporal dynamics of insect pests and to the best of our knowledge, there are few published studies on *B. tabaci* in soybean. Thus, the aim of this study was to determine the spatiotemporal distribution of adults and nymphs of *B. tabaci* in soybean crops from planting to harvest by using geostatistics and verifying abiotic factors (temperature, relative humidity, wind, and rainfall) that affect the dispersion of these insects.

2. Materials and methods

The experiment was carried out at Celeiro Seeds Farm, and the farm's total area was estimated at 16,000 hectares, where grains and soybean seeds are produced. The farm is located in the municipality of Monte Alegre do Piauí and Serra do Quilombo in the Brazilian state of Piaui (09°21′12"S; 45°07′42" W, 642 m) (**Figure 1A**). The region's soil is characterized as a yellow Latosol (Oxisol)28, and the climate is tropical with a dry season in winter, Aw, according to the Köppen-Geiger classification 29. Climatic data such as the mean air temperature (°C), relative humidity (%), wind speed (m/s) and wind direction were obtained from NIMET (National Institute

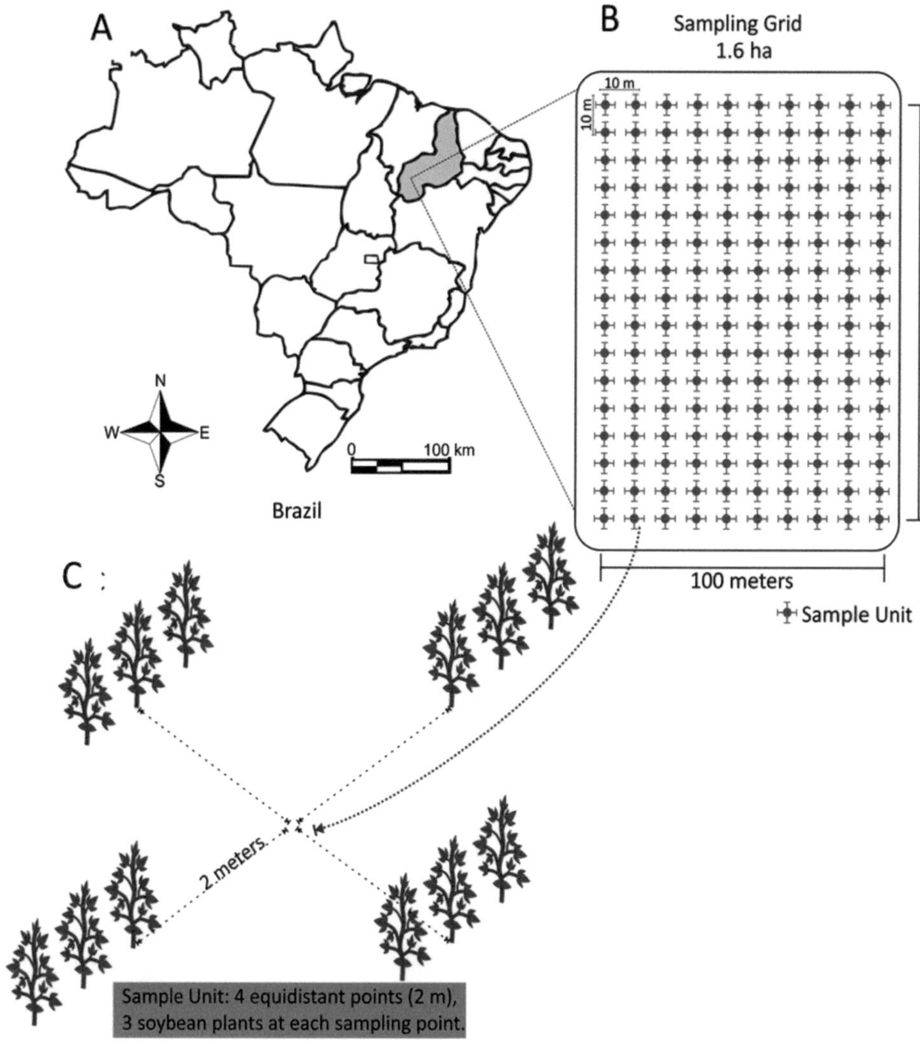

Figure 1.
(A) Brazil's map - Monte Alegre do Piauí, Piauí. (B) sample grid of the evaluated points and geographic location. (C) sampling unit (3 plants equidistant from the central point, total of 12 plants per unit).

of Meteorology), and rainfall data were obtained with a rainfall meter during the period of execution of the experiment (**Figure 2**). This work was carried out for 2 seasons (2017/2018 and 2018/2019). During the first season, the cultivars Brasmax BonoIPRO® and BRS 9180IPRO® were used; in the second season, the cultivars were Brasmax Bonus IPRO® and Brasmax Extreme IPRO®. The characteristics of the cultivars are shown in **Table 1**.

The area of 1.6 ha (16,000 m^2), with 100 m length and 160 m width, was divided into 160 sampling units, with 10 m^2. In the center of each sample unit, sets of four ordered points, equally spaced 2 m from a central point, were evaluated to improve the estimate of the nugget effect. Each sampling point was georeferenced with the aid of a GPS device model GPSMAP 60CS®x (Garmin). The distance from each field to the nearest native forest is described in **Table 1**.

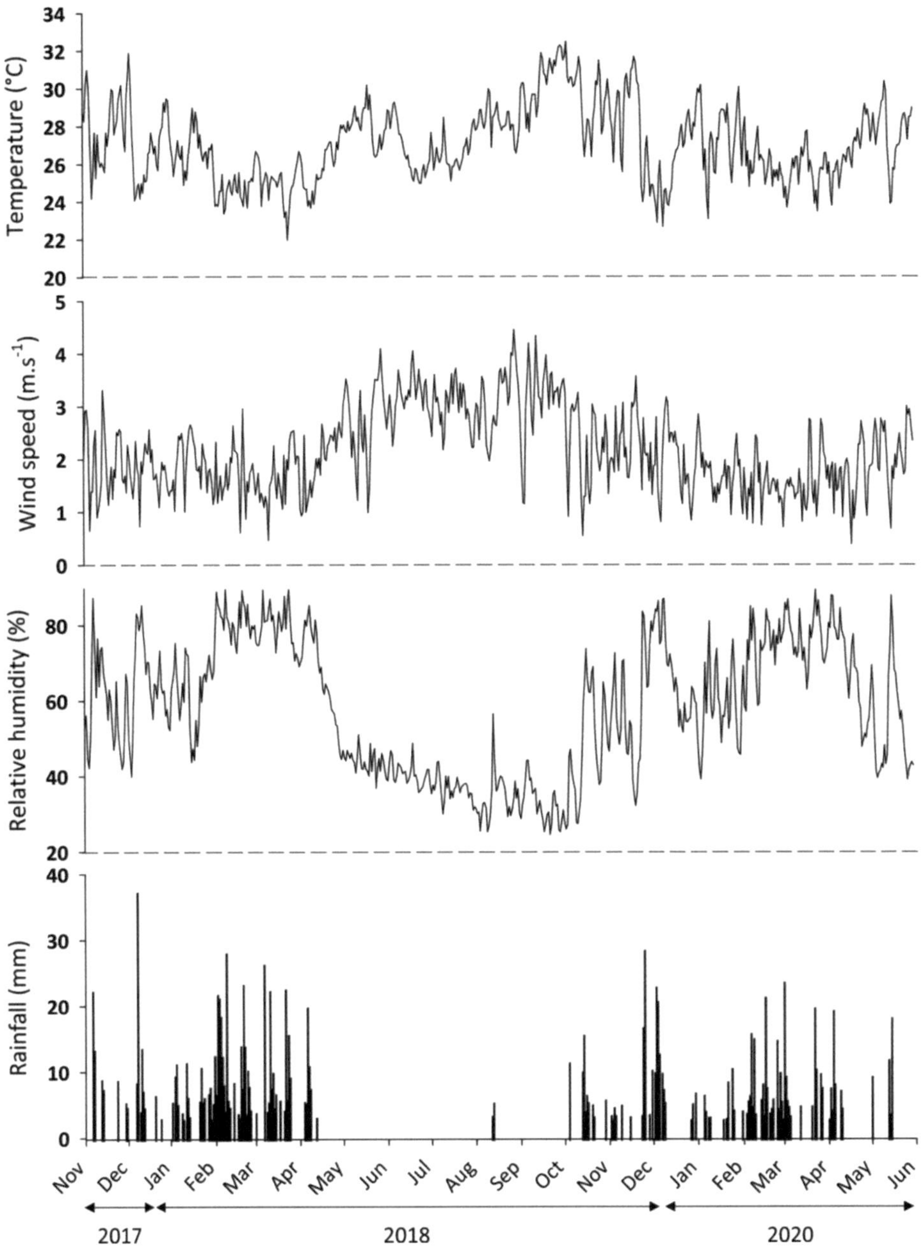

Figure 2.
*Rainfall, temperature and average relative humidity of the region. Celeiro seed farm in Serra do Quilombo, Monte Alegre do Piauí. National Institute of meteorology (INMET) data. * sowing month of cultivars.*

2.1 Evaluation of the *B. tabaci* density

The density of insects was evaluated in the sampling grid of 160 points previously georeferenced (**Figure 1B**). At each georeferenced point, 12 plants were evaluated; four subsamples of three plants were taken (sampling unit). The subsamples were 2.0 m equidistant from the central point (**Figure 1C**). The evaluated plants were

Season	I (November 2017–April 2018)		II (November 2018–April 2019)	
Variety	**Brasmax BonusIPRO ®**	**BRS 9180IPRO®**	**Brasmax BonusIPRO ®**	**Brasmax ExtremaIPRO ®**
Field	Field I	Field II	Field III	Field IV
Planting date and characteristics	Maturity group 7.9 - average cycle of 108 days. Planting on November 11, 2017	Maturation group 9.1 - average cycle 119–139 days. Planting on November 18, 2017.	Planting on November 27, 2018.	Maturation group 8.1 - average cycle of 110 days. Planting on December 4, 2018
Stand and spacing	390 thousand plants/ ha. Spacing between lines of 0.50 m	200 thousand plants/ ha. Spacing between lines of 0.45 m	390 thousand plants/ha. Spacing between lines of 0.50 m	200 thousand plants/ha Spacing between lines of 0.45 m
Fertilization	620 kg/ha of Simple SuperPhosphate, 200 kg/ha of Potassium Chloride, 3 kg/ha of copper sulphate, 2.5 kg/ ha of Manganese Monoxide, 3 kg/ha of Ulexite and 20 kg/ha of MIB Granary.	200 kg/ha Simple SuperPhosphate, 200 kg/ha of Potassium Chloride and 20 kg/ha of MIB Granary in the planting line.	630 kg/ha of Simple SuperPhosphate, 200 kg/ha of Potassium Chloride, 20 kg/ ha of Manganese Monoxide, 6 kg/ha of Ulexite and 13 kg/ha of Zinc Sulphate, by haul.	
Nearest native forest	1.79 km to the North	4.09 km to the West	4.22 km North	0.32 km Northwest
Assessments days	14, 28, 42, 56, 70 and 84 DAE[a]	17, 31, 45, 59, 73, 87 and 101 DAE.	13, 27, 41, 55, 69, 83 and 87 DAE	20, 34, 48, 62, 76, 90 and 104 DAE
Geographic coordinates	9°23′27.21"S 45° 6′56.62"W	9°25′3.54"S 45° 0′23.63"W	9°24′10.88"S 45° 8′37.50"W	9°22′25.27"S 45° 6′56.69"W

[a]= *Days After Emergence.*

Table 1.
Phenological cycle, fertilization, plant population, planting date, distance from cultivars to the nearest native forest, and evaluation days by cultivar in the two seasons studied.

positioned along a regular grid pattern throughout the crop cycle to obtain systematic sampling points and avoid directional trends.

The third leaflet from top to bottom (apical third) of each plant was evaluated by direct counting. Leaves were handled with care, and the nymphs and adults of *B. tabaci* were counted. A magnifying glass was used to count the number of nymphs. We evaluated these leaves in particular using a direct counting method because these are the ideal sample type and technique for assessing the density of *B. tabaci* nymphs and adults in soybean crops 14. The sampling schedule was standardized; the evaluations started at 7 am and ended at 11 am. The sampling schedules for each cultivar are described in **Table 1**.

2.2 Analysis of the spatial distribution of *B. tabaci* in soybean

The data on adult and nymph densities of *B. tabaci* were submitted to statistical analysis. Subsequently, principal component analysis was performed between the range and density of whitefly adults with climatic parameters (temperature, relative humidity, wind speed, and rainfall). These analyses were performed using R®

software. 30 All geostatistical procedures were performed using the Geostatistical Analyst Tool for ArcGIS 10.5 (ESRI). The procedure in this tool can be simplified as follows: map and examination of data; preprocess of data if necessary (transform, detrend, decluster); definition of spatial structure model; definition of search strategy; prediction of values at unsampled locations; quantification of uncertainty of the predictions; checking if the model produces reasonable results for predictions and uncertainties and, using the information in risk analysis and decision making.

Subsequently, geostatistical analysis was performed; the spatial patterns and interpolations were determined using the parameters of adjusted experimental semivariograms and ordinary kriging, respectively. The semivariograms were calculated from the primary collected data, allowing us to detect differences between pairs of sampled points in relation to the distances (this procedure was used to adjust the theoretical semivariogram). Once the semivariance increases, there is a spatial dependence relationship between the densities of *B. tabaci* and the sampled points. The nugget effect and the sill value were calculated for each of the adjusted models (spherical, exponential, and Gaussian).

The experimental semivariograms were adjusted to the theoretical models, and the selection was made based on the cross-validation parameters, in which the measured and estimated values were compared using the standard mean error (SME) and the root mean square standardized error (RMSSE).

The kriging indicator was applied to spatial distribution maps of the insect in the crops, with the objective of modeling the probability of unsampled locations exceeding the quality reference values (QRVs). The kriging indicator does not use normal distribution assumptions, because it transforms the original data into binary values and, consequently, into cumulative distribution functions from pre-established values, in this case the QRVs. 31, 32 The data were transformed into log-normal values to minimize distribution errors and meet the requirements of common kriging based on the rejection of the null hypothesis of the Kolmogorov-Smirnov test for normal distribution. Spatial variability was determined from isotropic and anisotropic semivariograms. Anisotropic calculations were performed in four directions (0, 45, 90 and 135°).

To obtain reliable estimates, the theoretical model needs to show SME values close to 0 and RMSSE close to 1.33 The Akaike information criterion (AIC) was used as the last selection criterion. The spatial dependence rate (SDR) was calculated according to the formula: $[C0/(C0 + C1) \times 100]$. 34 A nugget effect less than or equal to 25% of the plateau was considered strong. The value was considered moderate when it was between 25 and 75% and weak when it was above 75%. Variables that showed SDR less than one unit were not considered.

Spatial distribution maps of *B. tabaci* adults and nymphs were prepared for each crop. In these maps, the predominant direction of the winds is shown.

3. Results

According to the statistical analysis, the values of variance were within the range 0 to 7.52 and were predominantly positive, where the peak of the normality curve of the results was higher than the standard value, the distribution curves were classified with positive asymmetries, and the coefficient of variation values were predominantly greater than 30% (Online Resource 1). In the two seasons evaluated, infestation by adults and nymphs of *B. tabaci* generally started between 30 and 50 days after the emergence of the plants. The maximum population density of 10 adults per plant and approximately two nymphs per leaflet were recorded 101 days after plant emergence (Online Resource 1 and **Figure 3**).

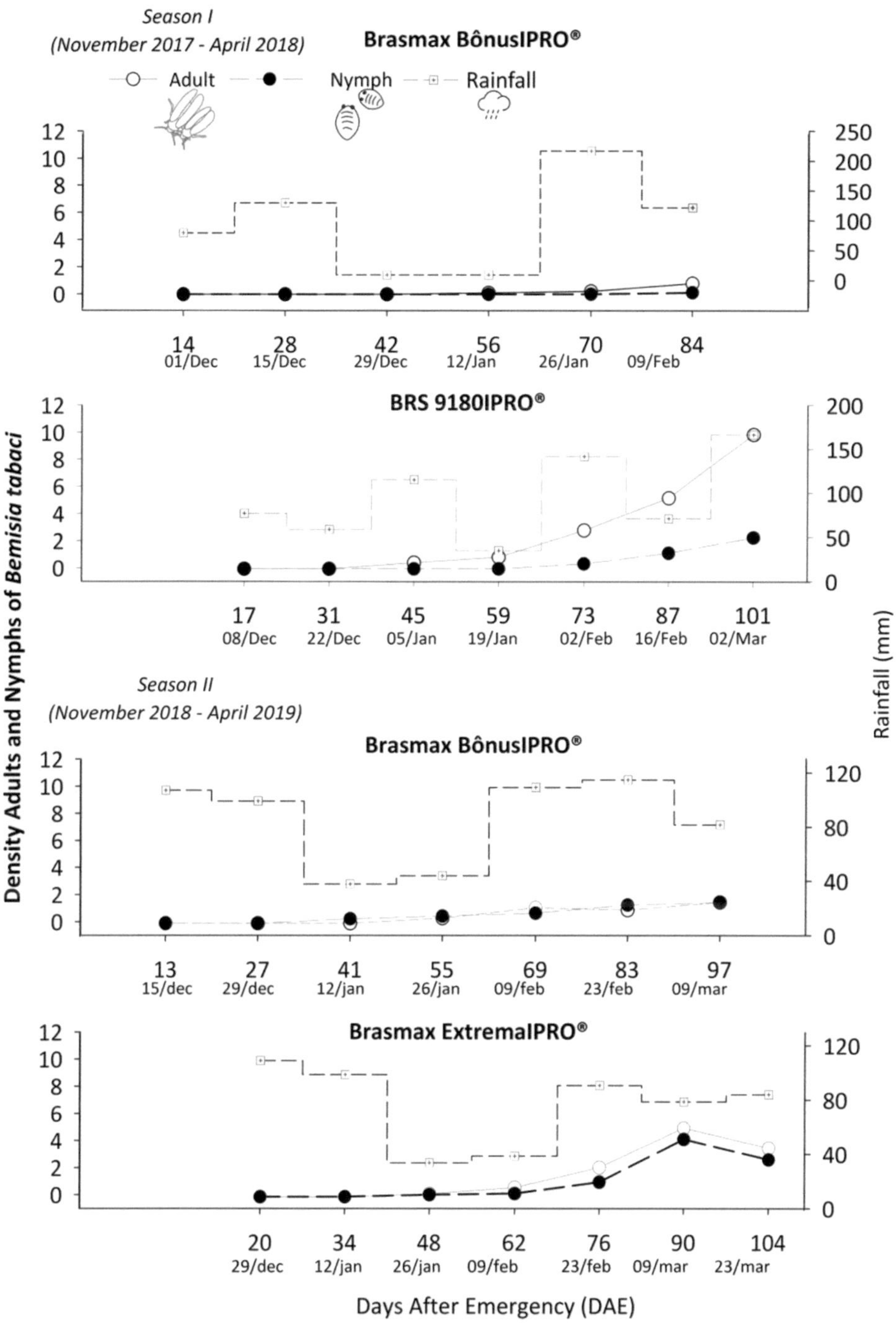

Figure 3.
Occurrence of Bemisia tabaci *adults and nymphs in the Brasmax BonoIPRO®, BRS 9180IPRO® and Brasmax ExtremaIPRO® cultivars for seasons I, II, and l.*

The results that showed a strong or moderate degree of spatial dependence (SDR) were subjected to geostatistical models and chosen based on cross-validation parameters using the standard mean error (SME) and the root mean square standardized error (RMSSE). Of the 126 models that were processed, 33 were selected, of which 13 were exponential, 10 Gaussian, 10 spherical, and 36 pure nugget effects (**Table 1**). All selected models were isotropic (i.e., the spatial autocorrelation was the same in all directions).

Differences in the variogram parameters (nugget, sill, and range) were observed in the three cultivars in the two sampled seasons. A nugget effect was observed in the variograms, always at a low density of *B. tabaci* adults or nymphs (**Figures 4** and **5**). The SDR of the models ranged from 0 to 100. This showed a significant effect of the nugget effect on the interpolation for some sampled days. These proportions displayed that the spatial component accounted for 66% of the total spatial variance. From the selected models, 15.15% demonstrated a strong SDR (<0.25), 51.52% moderate SDR (between 0.25 and 0.75), and 33.33 weak SDR (**Table 2**).

The ranges of the models varied from 13.03 to 132.53 m, and the maximum range obtained for adults was 67 m at 104 DAE in the Brasmax ExtremaIPRO® cultivar and approximately 130 m for nymphs 101 DAE in the BRS 9180IPRO® cultivar (**Table 2**).

We observed that colonization of soybean plants by *B. tabaci* may be divided into three stages: the start of infestation, colonization of the area, and dispersion in the area. The adults of *B. tabaci* began to infest the experimental area from the outermost area; in season I, the insects came from the west, and in season II, the insects originated from the north (**Figure 4**).

Colonization by *B. tabaci* nymphs occurred as adults disperse in the area. Near the end of the culture cycle, adults, and nymphs of *B. tabaci* had already colonized the whole experimental area. With the colonization of the total area, adults began the process of migration to nearby areas (**Figures 4** and **5**).

The soybean cultivars were divided into three groups according to the density of the pest. In the first group (Brasmax Bonus IPRO®), the lowest densities of the pest (2 adults and 1 nymph per leaf) were observed. In the second group (Brasmax ExtremaIPRO®), lands with a higher density of nymphs (up to 10 nymphs per leaf) were observed. In the third group (BRS 9180IPRO®), plants exhibited the highest densities of adults (up to 15 adults per leaf) (**Figures 4** and **5**).

We observed two patterns of variation in the densities of *B. tabaci* throughout the development of the plants. The first occurred in Brasmax Bonus IPRO®, where the pest population varied little over time and remained at low density. The second pattern occurred in Brasmax ExtremaIPRO® and BRS 9180IPRO®, where the pest population increased over time and was distributed throughout the crop area (**Figures 4** and **5**).

The mean air temperature was high (29–33°C) during the cropping period. On the other hand, rainfall ranged from 10 to 127 mm per month during the cropping period (**Figure 2**). The winds during the experimental periods occurred predominantly in the northeast direction in 70% of cases with a maximum speed of 2.1 m/s. Performing the main component analysis, it was apparent that the two axes explained 74.79% of the data variation. There were positive correlations between the range and wind speed, and this trend was observed also, in addition, between adult whitefly densities and rainfall and relative humidity (**Figure 6**).

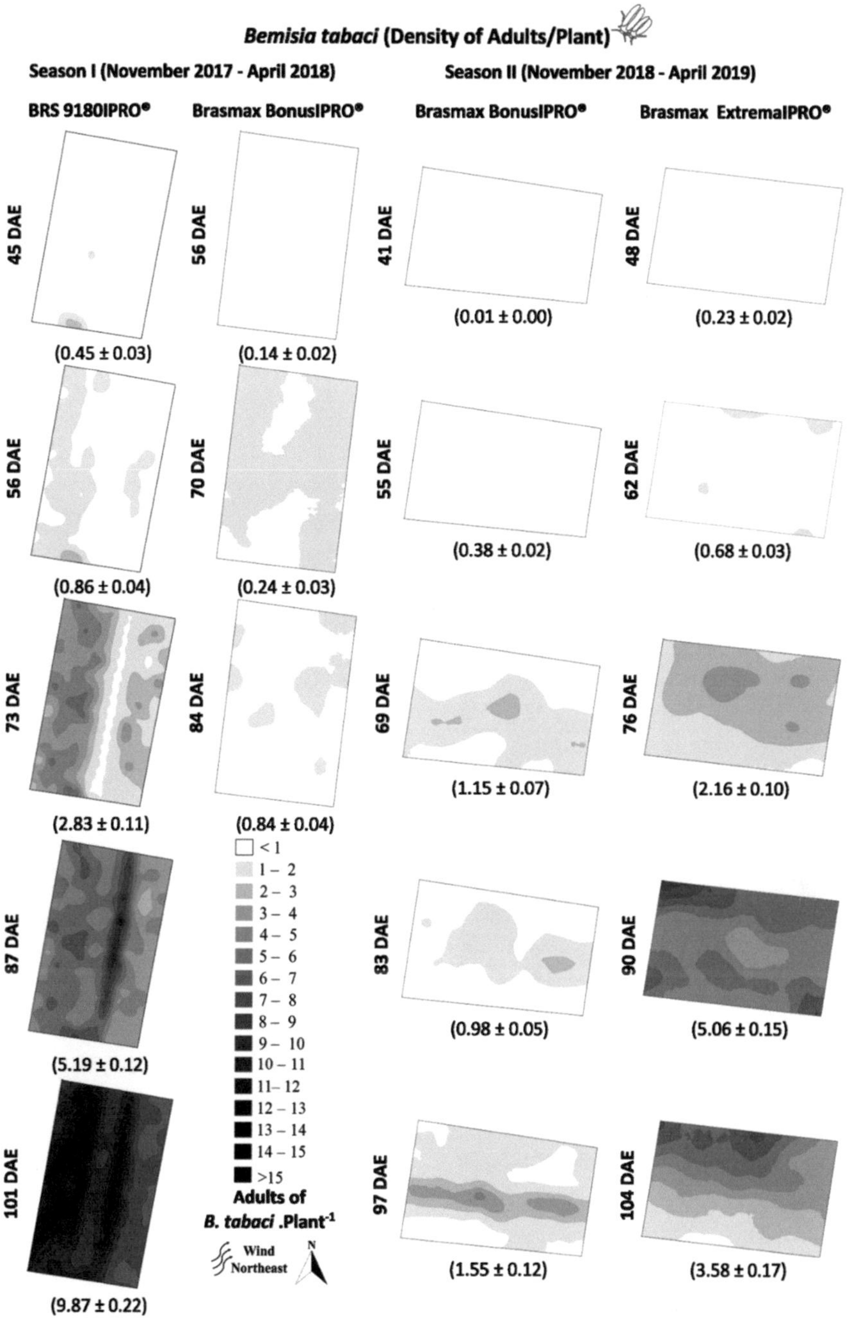

Figure 4.
Kriging maps for the spatial variability of Bemisia tabaci *adults in the Brasmax BonoIPRO®, BRS 9180IPRO® and Brasmax ExtremaIPRO® cultivars for the 2017/2018 season I and 2018/2019 season II.*

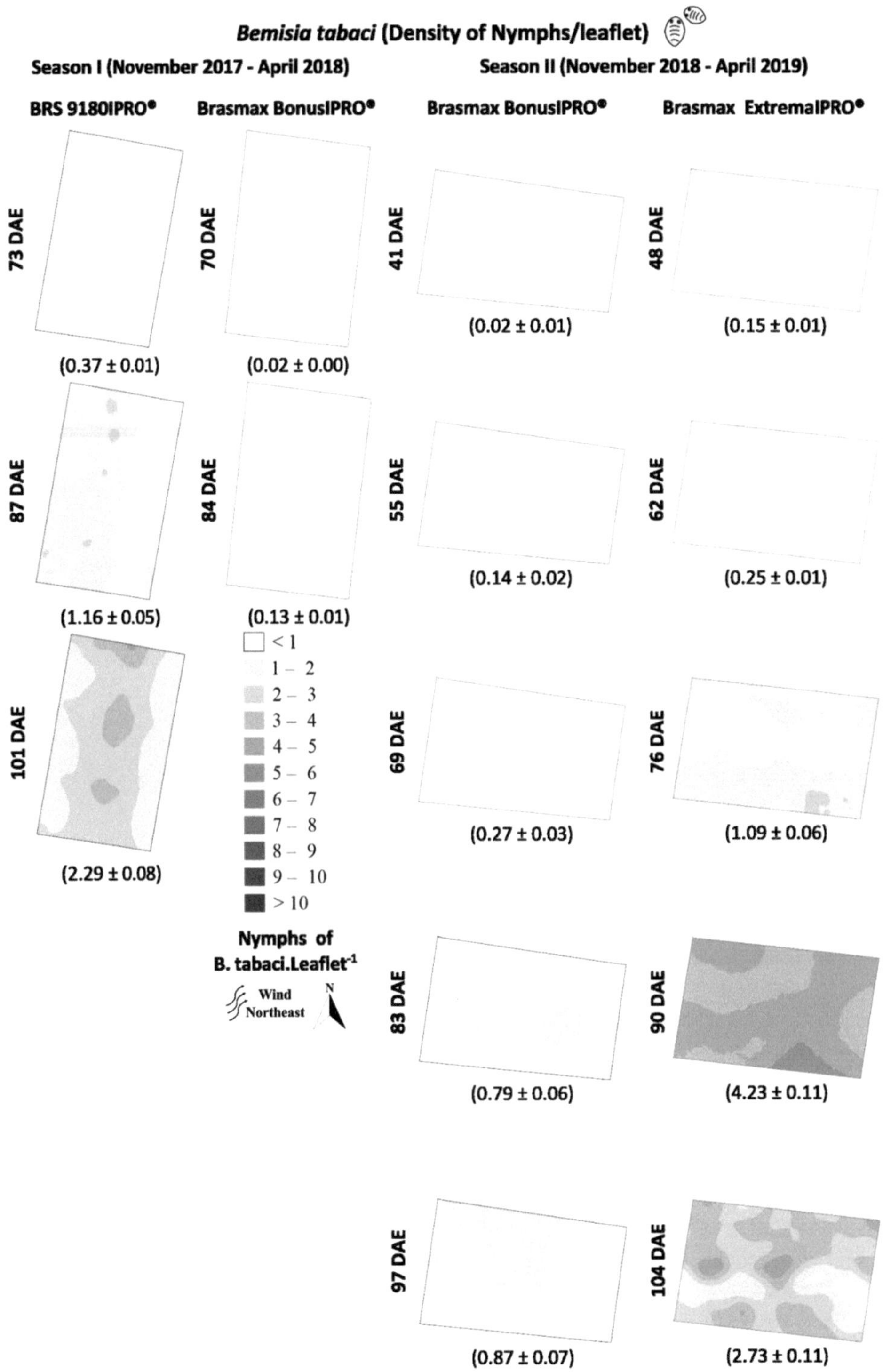

Figure 5.
Kriging maps referring to spatial variability of Bemisia tabaci *nymphs in the Brasmax BonoIPRO®, BRS 9180IPRO® and Brasmax ExtremaIPRO® cultivars for the 2017/2018 season I and 2018/2019 season II.*

Variety	Stages	SAD	Model	C_0 [a]	C_0+C_1 [b]	RDS (%) [c]	RDS	Range (m)	RMSSE [d]	SME and
Season I (November 2017–April 2018)										
Brasmax BonusIPRO ®	Adult	56	Exponential	0.06	0.08	72.02	Moderate	24.91	0.98	0.29
		70	Exponential	0.09	0.12	74.19	Moderate	24.91	0.98	0.35
		84	Exponential	0.14	0.24	60.06	Moderate	45.41	1.07	0.46
	Nynph	70	Exponential	0.00	0.00	97.83	Weak	34.40	1.00	0.05
		84	Exponential	0.00	0.02	0.00	Strong	27.01	0.96	0.11
BRS 9180IPRO®	Adult	45	Exponential	0.00	0.16	0.00	Strong	27.74	0.99	0.33
		59	Exponential	0.00	0.25	0.00	Strong	52.13	1.02	0.32
		73	Exponential	0.00	2.01	0.00	Strong	41.37	0.81	1.02
		87	Exponential	0.00	2.89	0.00	Strong	33.73	0.94	1.33
		101	Exponential	2.28	8.20	27.78	Moderate	52.59	0.98	2.29
	Nymph	73	Exponential	0.04	0.11	40.00	Moderate	20.50	0.96	0.33
		87	Exponential	0.12	0.42	27.87	Moderate	54.50	0.96	0.52
		101	Exponential	0.53	1.27	41.24	Moderate	132.53	0.99	0.86
Season II (November 2018–April 2019)										
Brasmax BonusIPRO ®	Adult	41	Spherical	0.00	0.00	100.0	Weak	72.82	1.06	0.03
		55	Spherical	0.04	0.06	70.0	Moderate	128.48	1.00	0.22
		69	Spherical	0.51	0.92	55.4	Moderate	45.18	0.96	0.83
		83	Spherical	0.28	0.53	52.8	Moderate	42.39	0.94	0.62
		97	Spherical	1.38	2.51	55.2	Moderate	25.20	0.89	1.48
	Nymph	41	Spherical	0.00	0.00	98.2	Weak	25.34	1.10	0.06
		55	Spherical	0.03	0.03	95.1	Weak	24.94	1.04	0.18
		69	Spherical	0.12	0.16	74.6	Moderate	20.82	0.96	0.40
		83	Spherical	0.54	0.55	96.5	Weak	45.18	0.94	0.76
		97	Spherical	0.73	0.73	100.0	Weak	128.48	0.94	0.88

Variety	Stages	SAD	Model	C_0[a]	C_0+C_1[b]	RDS (%)[c]	RDS	Range (m)	RMSSE[d]	SME and
Brasmax ExtremaIPRO ®	Adult	48	Gaussian	0.064997639	0.08	80.48	Weak	83.89	1.01	0.26
		62	Gaussian	0.108601774	0.17	65.00	Moderate	21.62	1.01	0.39
		76	Gaussian	1.167979313	1.75	66.74	Moderate	41.45	0.96	1.17
		90	Gaussian	2.574639589	3.34	77.07	Weak	24.10	0.99	1.79
		104	Gaussian	1.444284816	5.23	27.64	Moderate	67.53	1.02	1.32
	Nymph	48	Gaussian	0.01394965	0.02	87.38	Weak	13.03	0.95	0.13
		62	Gaussian	0.033389883	0.03	95.60	Weak	29.94	0.98	0.19
		76	Gaussian	0.167840405	0.41	41.16	Moderate	19.93	0.98	0.65
		90	Gaussian	1.675549944	2.05	81.93	Weak	35.23	0.99	1.38
		104	Gaussian	0.823629539	1.81	45.55	Moderate	22.01	0.96	1.18

[a] *Nugget Effect.*
[b] *Threshold.*
[c] *(C0/C0 + C1) * 100) = Spatial Dependence Rate.*
[d] *Root Mean Square Standardized Error; and: Standard Mean Error.*

Table 2.
Models and parameters estimated by semivariogram for Bemisia tabaci *adults and nymphs in the cultivars Brasmax BonoIPRO®, BRS 9180IPRO® and Brasmax ExtremaIPRO.*

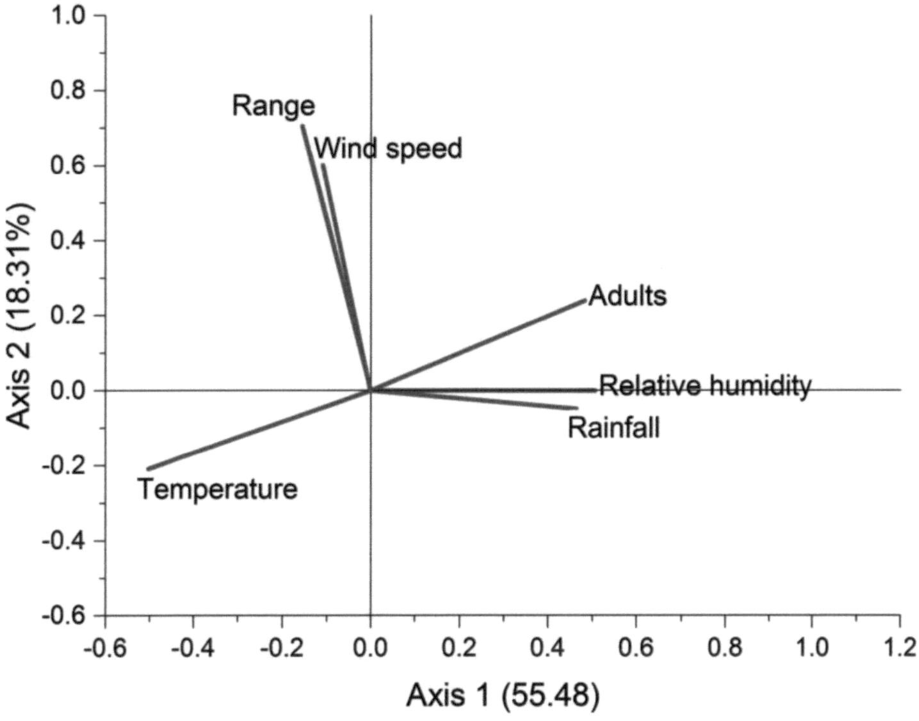

Figure 6.
Principal component analysis of the effects of rainfall and climatic elements on the range and abundance of Bemisia tabaci *in soybean crops. The first and second axes explained 74.79% of the variance.*

4. Discussion

The data presented here provide information on the spatiotemporal dynamics of *B. tabaci* in commercial soybean crops. Under the conditions evaluated, the initial foci of colonization by adults occurred near the end of the vegetative stage of the crop, and it is common to observe the occurrence of nymphs and adults of *B. tabaci* at this stage of plant development . The peak infestation by adults and nymphs recorded in the present study was relatively low compared to other studies [22, 28].

The cultivar Brasmax BonoIPRO® had high attractiveness for adults and nymphs of *B. tabaci*. However, there was low infestation in the two evaluated seasons, which may indicate that cultivars of an early cycle with early planting may escape the whitefly attack [29], as mentioned above. *B. tabaci* infestation was more severe in late February and early March.

B. tabaci showed an aggregated distribution in the soybean crops, indicating that there were factors that influenced this outcome. The spatial distribution of pest insects in crops is a consequence of the colonization and dispersion of these organisms at these sites [24, 30]. Climate elements, terrain topography, characteristics of insect species, and landscapes are among the factors influencing the spatial distribution of pest insects in crops [11, 24, 31–34].

Some results obtained for *B. tabaci* adults and nymphs showed a weak degree of spatial dependence. This finding is characterized as a pure nugget effect (total absence of spatial dependence), probably due to the low infestation by adults and whitefly nymphs in each evaluation, making the presence of insects in geostatistical analyses imperceptible [25, 35]. These results are a common finding in this type of study; other authors with experiments on the spatial distribution of *Euschistus heros* (Fabricius, 1798) (Hemiptera: Pentatomidae) in soybean have made similar observations [25, 36].

According to the semivariograms made, whitefly adults and nymphs had an isotropic distribution. This finding confirms the premise that *B. tabaci* disperses across the field in a diffusion pattern, allowing us to achieve an equal or similar number of adults throughout the sampling points regardless of direction, with the center of origin as the beginning of the evaluation [3]. Isotropic and aggregate pattern distributions were observed for adults and nymphs, and a reduction in range (dispersion radius) was observed as the density of adults per plant increased, while for nymphs, the opposite effect was verified. Other authors have observed the same pattern of whitefly distribution in soybean and other crops [3, 20, 22]. In experiments focused on the spatial distribution of *B. tabaci* in commercial watermelon crops, it was demonstrated that the dispersion of the insect was not influenced by a physical barrier or height gradient [20].

With the kriging maps, it was possible to verify that colonization of the area by *B. tabaci* proceeded in the western and northern regions. Near the experimental area, there was a permanent preservation area, native forest. Therefore, when favorable environmental conditions existed, the infestation of *B. tabaci* in soybean crops was to be expected. Migration from the native forest to the crop occurred when there was abundant food and shelter [26]. However, the northeast winds were predominant in both seasons; that is, the adults of *B. tabaci* flew against the wind searching for food. This fact is common since wind helps in the dispersion of volatiles in the area. Insects use volatiles as olfactory cues to find food, so insects can fly upwind, cross-wind and downwind when having perceived volatiles [37].

Kriging and semivariogram maps are fundamental tools that can assist in the identification and determination of the spatiotemporal dynamics of pests, in addition to providing information such as the dispersion pattern and range of insects. Based on this information, an adequate and efficient pest management monitoring program may be defined [20, 22, 25]. In addition, the elaboration of the maps allows the identification of the places with the highest incidence of whiteflies in crops, as well as the delimitation of the direction of colonization by adults. This allows the application of efficient control measures [20].

Geostatistics allowed us to verify the movements of *B. tabaci* adults and nymphs during the evaluations and soybean cycle. As previously explored, there are factors that can affect the distribution of pests in the field, features linked to insect characteristics such as population growth (reproduction, mortality) and dispersion (immigration, colonization, emigration) [3, 22, 25, 27, 36]. *B. tabaci* may fly to a height of up to 7 m and a distance of 7 km [38]. Climatic factors such as rainfall and the relative humidity of the air favored an increase in the density of whitefly adults, while wind speed favored the dispersion of adults, according to the literature [3, 20]. The development of monitoring and management techniques aimed at the precise control of *B. tabaci* depends on a better understanding of the flight behavior of the whitefly, especially short-range migration. The present study provides important information about the pattern of aggregation and distribution of *B. tabaci*, which may be used in future work to assist in the sampling method.

Acknowledgements

We thank the National Council for Scientific and Technological Development — CNPq — Brazil for financial support. We also thank the Coordination for the

Improvement of Higher Education Personnel (CAPES), Brazil and the Piauí State Research Foundation (FAPEPI) for the scholarships and resources provided. To the Federal University of Piauí, Brazil, for providing logistical support. We also thank the Celeiro Seeds Farm for the supply of the area and inputs.

Author details

Luciana Barboza Silva*, Raimundo Henrique Ferreira Rodrigues,
Thiago Ferreira Rodrigues, Maria Carolina Farias e Silva, Edivania de Araujo Lima and José Wellington Batista Lopes
Graduate Program in Agricultural Sciences, Federal University of Piauí,
Bom Jesus, Piauí, Brazil

*Address all correspondence to: lubarbosabio@ufpi.edu.br

References

[1] Lima CH, Sarmento RA, Pereira PS, et al. Feasible sampling plan for *Bemisia tabaci* control decision -making in watermelon fields. Pest Management Science. 2017;**73**:2345-2352

[2] Oliveira C, Auad A, Mendes S, et al. Economic impact of exotic insect pests in Brazilian agriculture. Journal of Applied Entomology. 2013;**137**:1-15

[3] Byrne DN, Rathman RJ, Orum TV, et al. Localized migration and dispersal by the sweet potato whitefly, *Bemisia tabaci*. Oecology. 1997;**105**:320-328

[4] Cruz P, Baldin E. Performance of *Bemisia tabaci* biotype B on soybean genotypes. Neotropical Entomology. 2017;**46**:210-215

[5] De Barro PJ, Liu SS, Boykin LM, et al. *Bemisia tabaci:* A statement of species status. Annual Review of Entomology. 2011;**56**:1-19

[6] Li JM, Ruan YM, Li FF, et al. Gene expression profiling of the whitefly (*Bemisia tabaci*) Middle East – Asia minor 1 feeding on healthy and tomato yellow leaf curl China virus -infected tobacco. Insect Science. 2011;**18**:11-22

[7] Stansly PAN, Naranjo SE, Brown JK, et al. *Bemisia*: Bionomics and Management of a Global Pest. Vol. 1. Netherland: Springer; 2010. p. 540

[8] Stansly PAN, Steven E. *Bemisia:* Bionomics and Management of a Global Pest. Vol. 1. Dordrecht, Springer Netherlands: Imprint, Springer; 2010. p. 558. ISBN: 978-90-481-2459-6

[9] Masuda K, Kato MT, Saito T. Reduction in carotenoid and chlorophyll content induced by the sweet potato whitefly, *Bemisia tabaci*. Scientia Horticulture. 2016;**200**:102-104

[10] Marubayashi JM, Yuki VA, Wutke EB. Transmission of cowpea mild mottle virus by the whitefly *Bemisia tabaci* biotype B to bean and soybean plants. Summa Phytopathologica. 2010;**36**:158-160

[11] Czepak C, Coelho ASG, Rezende JM, et al. *Bemisia taba* ci MEAM 1 population surveys in soybean cultivation. Entomology Experimentalis and Applicata. 2018;**166**:215-223

[12] Da Silva Oliveira CE, Carneiro DEF, Toscano LC, et al. Population dynamics of *Bemisia tabaci* biotype B (Gennadius, 1889) in transgenic soybean. Revista de Agricultura Neotropical. 2018;**5**:1-5

[13] Vieira S, Bueno A, Boff M, et al. Resistance of soybean genotypes to *Bemisia tabaci* (Genn.) biotype B (Hemiptera: Aleyrodidae). Neotropical Entomology. 2011;**40**:117-122

[14] Ellsworth PC, Martinez-Carrillo JL. IPM for *Bemisia tabaci:* A case study from North America. Crop Protection. 2001;**20**:853-869

[15] Ellsworth PC, Diehl JW, Dennehy TJ, et al. Sampling Sweet Potato Whiteflies in Cotton. Tucson, AZ, United States: College of Agriculture and Life Sciences, Cooperative Extension IPM Series, University of Arizona; 1995. p. 2

[16] Naranjo SE, Flint HM. Spatial distribution of adult *Bemisia tabaci* (Homoptera: Aleyrodidae) in cotton and development and validation of fixed-precision sampling plans for estimating population density. Environmental Entomology. 1995;**24**:261-270

[17] Palumbo J, Tonhasca AD Jr, Byrne D. Sampling Plans and Action Thresholds for Whiteflies on Spring Melons. Online. Coop. Ext. IPM Series 1. Publ. No. 194021. Tucson, Arizona: University of Arizona, College of Agriculture and Life Sciences, Cooperative Extension; 1994 Available from: http://cals.arizona.edu/crops/vegetables/insects/wf/wfmelons.html

[18] Moura MFD, Picanço MC, Silva ÉD, et al. Sampling plan of the biotype B of *Bemisia tabaci* in the cucumber crop. Pesquisa Agropecuária Brasileira. 2003;**38**:1357-1363

[19] Galdino TVS, Ferreira DO, Santana Júnior PA, et al. The role of the beetle *Hypocryphalus mangiferae* (Coleoptera: Curculionidae) in the spatiotemporal dynamics of mango wilt. Journal of Economic Entomology. 2017;**110**:865-874

[20] Lima CH, Sarmento RA, Galdino TV, et al. Spatiotemporal dynamics of whitefly *Bemisia tabaci* (Hemiptera: Aleyrodidae) in commercial watermelon crops. Journal of Economic Entomology. 2018;**111**:1895-1903

[21] Naveen N, Chaubey R, Kumar D, et al. Insecticide resistance status in the whitefly, *Bemisia tabaci* genetic groups Asia-I, Asia-II-1 and Asia-II-7 on the Indian subcontinent. Scientific Reports. 2017;7:1-15

[22] Suekane R, Degrande PE, Melo EPD, et al. Spatial distribution of soybean plants infested with whitefly Bemisia tabaci (Gennadius, 1889) (Hemiptera: Aleyrodidae). Archives of the Biological Institute. 2018;**85**:1-6

[23] Felicio TNP, Costa TL, Sarmento RA, et al. Surrounding vegetation, climatic elements, and predators affect the spatial dynamics of *Bemisia tabaci* (Hemiptera: Aleyrodidae) in commercial melon fields. Journal of Economic Entomology. 2019;**112**:2774-2781

[24] Martins JC, Picanço MC, Silva RS, et al. Assessing the spatial distribution of *Tuta absoluta* (Lepidoptera: Gelechiidae) eggs in open -field tomato cultivation through geostatistical analysis. Pest Management Science. 2018;**74**:30-36

[25] Pias OHDC, Cherubin MR, Berghetti J, et al. Sampling grids for monitoring *Euschistus heros* infestation in soybean. Bragantia. 2017;**76**:386-397

[26] Macfadyen S, Hopkinson J, Parry H, et al. Early-season movement dynamics of phytophagous pest and natural enemies across a native vegetation-crop ecotone. Agriculture, Ecosystems & Environment. 2015;**200**:110-118

[27] Sciarretta A, Trematerra P. Geostatistical tools for the study of insect spatial distribution: Practical implications in the integrated management of orchard and vineyard pests. Plant Protection Science. 2014;**50**:97-110

[28] Vieira SS, Bueno RCODF, Bueno ADF, et al. Different timing of whitefly control and soybean yield. Ciência Rural. 2013;**43**:247-253

[29] Ameren N, Amjad M, Murtaza B, et al. Biogeochemical behavior of nickel under different abiotic stresses: Toxicity and detoxification mechanisms in plants. Environmental Science and Pollution Research. 2019;**26**:10496-10514

[30] Rosado JF, Picanço MC, Sarmento RA, et al. Geostatistics as a tool to study mite dispersion in physic nut plantations. Bulletin of Entomological Research. 2015;**105**:381-389

[31] Carrière Y, Degain B, Liesner L, et al. Validation of a landscape-based model for whitefly spread of the cucurbit yellow stunting disorder virus to fall melons.

Journal of Economic Entomology. 2017;**110**:2002-2009

[32] Ludwig M, Schlinkert H, Meyhöfer R. Wind -modulated landscape effects on colonization of B Russels sprouts by insect pests and their syrphid antagonists. Agricultural and Forest Entomology. 2018;**20**:141-149

[33] Ruscha A, Valantin-Morisona M, Sarthou JP, et al. Effect of crop management and landscape context on insect pest populations and crop damage. Agriculture, Ecosystems & Environment. 2013;**166**:118-112

[34] Silva GA, Picanço MC, Bacci L, et al. Control failure likelihood and spatial dependence of insecticide resistance in the tomato pinworm, *Tuta absoluta*. Pest Management Science. 2011;**67**:913-920

[35] Reay-Jones FPF, Toews MD, Greene JK, et al. Spatial dynamics of *stink bugs* (Hemiptera: Pentatomidae) and associated boll injury in southeastern cotton fields. Environmental Entomology. 2010;**39**:956-969

[36] De Souza LA, Barbosa JC, Aguirre-Gil OJ, et al. Spatial distribution of nymphs and adults of *Euschistus heros* (Fabricius, 1794) (Heteroptera: Pentatomidae) in transgenic soybean cultivars of different maturing cycles. Semina: Agricultural Sciences. 2018;**39**:905-920

[37] Reinecke A, Hilker M. Plant semiochemicals – Perception and behavioral responses by insects. Annual Plant Reviews online. 2014;**47**:115-153

[38] Isaacs R, Byrne DN. Aerial distribution, flight behavior and egg load: Their inter -relationship during dispersal by the sweetpotato whitefly. Journal of Animal Ecology. 1998;**67**:741-750

Section 2

Food Detection and Feeding Behavior

Chapter 3

Food Detection and Feeding Behavior of Three Species of Household Cockroaches, *Blatella germanica* (L.), *Periplaneta americana* (L.), and *Supella longipalpa* (F.)

Anil Chandra Neupane

Abstract

German cockroaches (*Blatella germanica* L.), American cockroaches (*Periplaneta americana* L.,) and brown-banded cockroaches (*Supella longipapla* F.) are the most important urban insect pests. The food detection and feeding behavior of these cockroaches are varied and depend on different factors. German cockroach starts feeding between 7:00–10:00 pm and 4:00–5:00 am, whereas the American cockroach starts within the first few hours of darkness followed by an inactive period in the latter part, and throughout the light period. The calling in females of brown-banded cockroaches followed periodicity and peak calling occurs in the scotophase. Likewise, the behavioral response of male brown-banded cockroaches was at a peak in the scotophase. German cockroach compound eye is sensitive to blue-green portion of the spectrum (major) and ultraviolet (UV) (minor). The compound eye of the American cockroach received the blue-green and violet (or ultraviolet) regions of the spectrum. Information on the compound eye sensitivity of brown-banded cockroaches is limited. The possession of specific hygroreceptors could play an important role in both German and American cockroaches. The German cockroach preferred carbohydrates food and consumed more containing starch, glucose, sucrose, mannitol, maltose, sorbitol, or glycerol. Very limited studies were carried out to determine the food detection ability and the feeding behavior of the brown-banded cockroach. Future studies should be directed toward the color preferences of the brown-banded cockroach.

Keywords: feeding behavior, preference, distilled grain, time budgeting, attractive chemicals

IntechOpen

1. Introduction

German cockroaches (*Blatella germanica* L.) (Dictyoptera: Blattellidae), American cockroaches (*Periplaneta americana* L.) (Dictyoptera: Blattidae), and brown-banded cockroaches (*Supella Longipalpa* F.) (Dictyoptera: Blattellidae) are the most important urban insect pests, cause nuisance in and around the house. Cockroaches, their body parts, saliva, and protein produced by them cause allergies and severe asthma as well [1–3]. The German cockroach is the most abundant and has the widest distribution. The brown-banded and American cockroaches follow the German cockroach with respect to importance in cosmopolitan distribution [4].

All three species acted as vectors to transmit pathogenic bacteria [5–10]. They are nocturnal in habit; however, their peak activity varies during scotophase. Wang and Bennett [11] found that the diet history of cockroach strains influenced the feeding behavior of German cockroaches. They further stated that cockroaches grown on a mixed diet consumed less gel bait than those reared only on a rodent diet. Insecticide bait formulations have been used effectively used for managing German cockroach populations [12, 13]. A similar principle has been used for the other two species of household cockroaches. In recent years, gel bait formulation has been highly effective in controlling German cockroach populations [14]. These baits are toxic and normally contain a sugar phagostimulant, as well as active ingredients that are incorporated into the food matrix [15]. The major influencing factors for bait efficacy are the attractiveness [16], palatability, and toxicity of the active ingredients [17]. It has been stated that field-collected German cockroaches develop an aversion to glucose [18] and subsequently reject the food containing this compound [19].

Field-collected gel bait-resistant Cincy strain of German cockroaches showed resistance with Avert (0.05% abamectin), maxforce FC (0.01% fipronil), and pre-empt (2.15% imidacloprid) gel baits [20, 21]. Wang et al. [20] further mentioned that the aversion Cincy strain exhibited avoidance behavior on agar gel bait that contained fructose, glucose, maltose, and sucrose, which were commonly used as phagostimulants. Modification of those inert ingredients improved the efficacy of fipronil gel. Milio et al. [22] stated that 1.65% hydramethylnon bait stations provided moderate control of American cockroaches in poultry house feed rooms. However, Holbrook et al. [23] found that German cockroaches showed resistance to fipronil even though their ancestors were never exposed to it and showed resistance to the cyclodienes, which were formerly used for cockroach control. Similarly, it had been reported that field-collected bait resistant (Cincy) strain of German cockroaches showed behavioral resistance to avert (0.05% abamectin), maxforce FC (0.01% fipronil) [20] and avert (0.05% abamectin), maxforce FC (0.01% fipronil), and pre-empt (2.15% imidacloprid) gel baits [21]. Moreover, both maxforce FC containing fipronil and hydramethylnon were not effective for the control of bait aversion (Miami) strain of German cockroaches [24]. Similarly, in Taiwan, German cockroaches collected from households and hospitals in Kaohsiung area, south Taiwan develop resistance against three insecticides propoxur, chlorpyrifos, and cypermethrin [25]. Agrawal et al. [26] found that the synthetic pyrethroid combination of imiprothrin 0.07% + cypermethrin 0.2% aerosol caused only 20% reduction of German and American cockroach infestation 12 weeks after treatment. Recently, Fardisi et al. [27] stated that both insecticide-susceptible lab strains and field strains of German cockroaches exhibited a varying level of resistance to indoxacarb, fipronil, acetamiprid, beta-cyfluthrin, bifenthrin, and lambda-cyhalothrin. Hydramethylnon and imidacloprid insecticides could not achieve more than 90% mortality for the field strain, although there was no survivorship difference between the field strains and lab strains with the two insecticides. Field strains showed the lowest resistance to boric acid, abamectin, dinotefuran, clothianidin, thiamethoxam, and chlorfenapyr. Liang et al. [28] identified that prolonged exposure to the German cockroach showed physiological resistance to baits containing fipronil or indoxacarb. They further determined that cockroaches exposed to fipronil-containing gel bait exhibited cross-resistance to indoxacarb-containing gel bait.

In recent years, most of the studies have been focused on the development of new gel bait and efficacy tests of gel baits. However, limited studies have been carried out to determine attractants food, food detection, and feeding behavior of household cockroaches. Thus, the study was done to collect and discuss information on food detection and feeding behavior of the German cockroach, American cockroach, and brown-banded cockroach.

2. Circadian rhythm among three cockroach species

Generally, cockroaches are nocturnal in habit. Previous researchers have found that German cockroaches, remain in a temporary or permanent sanctuary or refuge for pests to live and or rest (harborage) during the day or other periods of light and become active during the night. They obtained a mixed diet from their foraging and are generally omnivorous [13, 29]. However, Tsai and Lee [30] determined that mated females could not express a circadian locomotor rhythm. In another study, it was found that the German cockroaches exhibited a bimodal activity response when they were exposed to natural light conditions and the activity began at sunset and reached a peak several hours after the onset of darkness [31]. Moreover, it was also found that a second peak activity occurred shortly before the light period began and was more noticeable under natural light than under artificial light [31].

In German cockroaches, daily patterns of activity generally varied between 22 and 26 hours [13]. External cues synchronize the activity periods and various periodic events typically serve as cues to entrain behavioral rhythms [13]. Adult male German cockroaches are more active than nymphs and adult females when foraging for food, water, and nest, as well as searching for mates [31]. The daily rhythm of male German cockroaches followed two phases, scotophase and photophase. German cockroach feeding behavior occurred between 7:00–10:00 pm and 4:00–5:00 am. The feeding behavior of German cockroaches involved swinging the antenna followed by touching the food [32]. Their activities increased rapidly at the onset of the scotophase from 14:00 hours to a peak at 17:00 hours and decreased rapidly afterward. Shortly before the photophase commences, there was a second rush of activity between 22:00 hours and 23:00 hours [31]. Furthermore, Fuchs [33] determined that the activity observed from German cockroaches peaked 3 h before the scotophase and 1 h before the photophase only when they were in the nest. Again, activity reached its highest peak immediately after the onset of the scotophase. He further revealed that long-range foraging occurred at specific times, whereas short-range foraging is common behavior of German cockroaches. In the case of females, their activity decreased to a minimum and there was no significant difference in activity during scotophase and photophase. The periods of maximal activity of the gravid females were often during photophase [31].

Sommer [34] examined substrate vibrations and the presence of other German cockroaches as potential signal and determined that the activity of cockroaches and individual rhythm could be fully synchronized with vibrations (50 Hz for 12 hours).

Moreover, he found that cockroaches dragged by photoperiod could not be synchronized with such 50 Hz vibrations.

Silverman [35] stated that the frequency of feeding and drinking of adult German cockroaches was determined by the distance from the nest to resource. Moreover, the cockroaches used a step-by-step manner to explore food patches, first, they search for food nearby the nest then they will move farther after the depletion of the nearest food [36]. Moreover, Nalyanya et al. [37] stated that both adult males and first instar German cockroaches chose shelters that were nearby the attractant-treated area and the furthest distance from the repellent-treated area. They further mentioned that the cockroach consumed the highest amount of food, and mortality was also the highest when the insecticide bait was placed near the preferred shelter.

In the case of American cockroaches, the onset of activities occurred within the first few hours of darkness, followed by a relatively inactive period in the later part and throughout most of the light period. However, adult females did not exhibit activity rhythms related to the lighting regimen (a specific photoperiod-12 L: 12 D h) [38]. In another experiment, it was determined that American cockroaches had diurnal locomotory activity correlated with alternating light and dark period; however, there was no correlation with temperature and humidity fluctuation. Alternate period of 24 h light and dark maintained a similar rhythm and activity which coincided with the beginning of day or night [39].

In the American cockroach, Lipton and Sutherland [38] demonstrated that the onset of feeding occurs soon after the onset of darkness. Moreover, Harker [40] found that the ocelli of American cockroaches were linked with light and darkness and also directly correlated with the establishment of the rhythm. Loss of rhythm was found when painting over the eyes and ocelli of the cockroaches. There was no direct connection between the hunger cycle and the rhythm activities.

The brown-banded cockroach, calling occurred discontinuously and followed a diel pattern primarily during the scotophase, in a 12 L: 12 h D photoperiod and after transfer to continuous light or dark conditions [41].

3. Food searching behavior

Ebeling et al. [42] reported that there was an inverse relationship between population density and exploratory behavior in the male German cockroach, which was accredited by the presence of an aggregation pheromone in the nest. Raubenheimer and Jones [29] mentioned that the German cockroach can balance the micronutrient intake by selecting a mixed diet from foods. The authors further stated that the ability to distinguish food of different nutritional values could be due to specific nutrient learning that was developed slowly on nutritionally imbalanced food. In general, when there are imbalanced foods, it shows slowed development and increased mortality in herbivores; however, the German cockroaches show well adapted variation of ingested nutrients. Eating and drinking activities of *B. germanica* were related to the more general circadian activity phases [31].

Similarly, Gadd and Raubenheimer [43] reported that American cockroaches were able to associate food odors with proteins. Moreover, it was determined that this species exhibited a spontaneous initial preference for vanilla over menthol [44].

Nymphs of brown-banded cockroaches self-selected a 15.5:84.5 protein: carbohydrate diet when they were reared with two imbalanced diet cubes. One contained absence of protein, while the other contained the absence of carbohydrates [45].

4. Eating and drinking behavior of cockroaches

The German cockroach is nocturnal in habit and leaves its shelter for foraging at night. The development stage of individual cockroaches is determined by their eating and drinking habits [46]. The population of *B. germanica* is distributed in contiguous patterns and forms aggregates. There are no differences between males and non-gravid females, but gravid females stayed mostly in the shelter and were less mobile [47]. Durbin and Cochran [48] observed that mortality increased when there was a deprivation of both food and water, which ultimately delayed the reproductive cycle. They reduced oothecal hatch in German cockroaches. Females survived much longer without food. Moreover, time spent on food and water was not much different for males and gravid females. Males foraged for food or water at least once in 7 days out of 10 days, whereas gravid females foraged actively on fewer than three days. With their fat and water reserves, female German cockroaches are well-suited to survive without food on a temporary basis [48].

The increased need for water also helped explain the behavior of most cockroaches to frequent water sources rather than solid food sources. The possession of specific hygroreceptors could play an important role in American cockroaches [31].

In the absence of drinking water, the nymphs of brown-banded cockroaches survived significantly longer than German cockroach nymphs. The capability of brown-banded nymphs was higher than German cockroach nymphs for producing and utilizing extra metabolic water from food [49]. Moreover, female, brown-banded cockroaches, when fed a 5% protein diet consumed less than fed either 25% protein or commercial rat food. Adult cockroach performance was also directly correlated with dietary protein levels and those females who died rapidly and were fed 65% protein [50].

5. Effect of population density on feeding and drinking behavior

It was found that male German cockroaches drink more at the low population density, although there was no overall effect of population density on feeding or drinking events [33].

Silverman [35] observed that when resources were farther away from the shelter the number of drinking and eating bouts increased. It was higher and took longer than at lower German cockroach population densities. If food and water were close to the shelter, shorter and fewer drinking bouts occurred under crowded conditions. This could be a reaction to increased competition at the water source and could relate to the interruption of drinking under high-density conditions. Under the crowded condition, more foraging occurred even if the resource was placed far from the harborage [35].

6. Factors affecting foraging and feeding behaviors

The age of German cockroaches was also influenced by the foraging behavior; adults and large instar (fifth and sixth instar) searched and found food sources first before the small instars [51]. Ballard et al. [52] determined that male German cockroaches were more active and explored more than females. Cochran [53] identified that feeding and drinking activity peaked during the egg maturation period and ended sharply at the appearance of egg capsules. However, Silverman [35] mentioned that nongravid female and male German cockroaches fed and drank more often than gravid females.

Cockroach feeding was affected by sex, age, and reproductive condition. The feeding behaviors of female *B. germanica* differed from nymph and adult males [54]. Males and unmated females of German cockroaches often responded to food odors from short distances, and they ate about 1–3 mg of food per day. Males who copulated twice a week ate more and died earlier than males that mated only once a week. In the case of females, different stages of reproductive cycles, carry different nutritional demands. Females, which were fed a low-protein diet, were able to increase consumption for supporting their normal growth and reproduction, whereas those fed with a high-protein diet increased mortality and decreased consumption and reproduction [55]. Peak feeding and drinking occurred during the egg maturation period and sharply ended when the egg case was apparent. Females feed and drink sparingly during the egg case-carrying period [53].

7. Visual sensitivity attractant color

Low-intensity red light is used to observe foraging and mating behavior of German cockroaches [56]. German cockroaches saw light with their compound eyes and received light through ocelli and dermal photoreceptors [57]. Koehler et al. [56] determined that German cockroaches had a color vision and the dorsal part of the compound eye had two peak sensitivities. The main visual wavelength spectrum was in the blue-green portion of the spectrum (490 nm), while the minor one was in the ultraviolet (UV) (365 nm). The UV light stimulated the highest level of locomotion, while green light stimulated about 30% and gold and red light did not affect locomotion.

Walther [58] mentioned that the eyes of *P. americana* contained at least two kinds of receptors. They were stimulated by the blue-green and violet (or ultraviolet) regions of the spectrum. Cockroaches contain compound eyes, which bear several kinds of receptor stimulated between 316 and 704 nm [59]. The spectral sensitivities of the dorsal ocelli of American cockroaches were measured using electrophysiological methods. Renowned showed that the waveform of the electrical response (ERG) of dark-adapted American cockroach ocellus was dependent on the intensity, but not on the wavelength of stimulating light. Moreover, they determined that the cockroach ocellus appeared to possess a single photoreceptor type, maximally sensitive at about 500 mμ [60].

8. Food attractant

Blattella germanica is oriented to food or water only if they came close to it after they began foraging, particularly after being deprived of food or water for a few days [31].

Cockroaches survived for long periods of time without food or water. *B. germanica* lived significantly longer if they have only food with access to water [31]. Willis and Lewis [61] reported that at 40% RH, without food and water, *B. germanica* survived up to 8 days, whereas females survived up to 13 days. Similarly, at 70% RH, females can survive up to 28 days without food or water. Water was required more critically than food for German cockroaches. Male *B. germanica* could not survive more than 9–10 days, but females could survive with access to water and without food for up to 45 days.

It was reported that after a carbohydrate meal, enzymes digested the sugars and the crops cleared within 48 hours. Sugar helped to increase the number of feeding episodes because of efficient digestion and the emptying of carbohydrates from the crop. Since German cockroaches preferred containing carbohydrates, bait consumption also increased by including substances, such as starch, glucose, sucrose, mannitol, maltose, sorbitol, or glycerol [4].

Karimifar [62] determined that peanut butter and beer were the most promising food source having semiochemicals that mediated the attraction of German cockroach. The molting and reproduction behavior of German cockroaches is regulated by food [46]. Food also played an important role in the temporal and spatial distribution of population densities of the German cockroach [36, 47].

Nalyanya et al. [16] found that bread was second to Maxforce gel for attracting male German cockroaches in olfactory assays. It was stated that in trapping studies in apartments, bread was used frequently and often mixed with beer [63–65]. It was identified that German cockroach baits are the mixture of odor or pheromone lures that attract insects [66].

Ballard and Gold [67] tested German cockroach attractiveness toward white bread, M. Sticky Chrysalis powder, Mr. Sticky Roach Bait, no bait, boiled raisins, potato, apple, dry dog food, German cockroach feces, German cockroach, dry yeast, banana, and Osage orange and found that the cockroaches were attracted significantly toward white bread and Osage orange.

Lofgren and Burden [68] stated that two percent Dipterex bait and powdered sugar showed the greatest attractiveness toward German cockroaches. Dextrin and corn starch alone or in combination were the most attractive to nymphs of American cockroaches in laboratory conditions. Similarly, it was found that German and American cockroaches were attracted to common foodstuffs, such as soft drink syrups, brown sugar, molasses, and essential oils of banana, sweet orange, apple, and pineapple [4]. Previously, insecticides were mixed with attractive food, such as honey, sugar, banana, beer, bread, cornmeal flour, potato, and peanut butter, that were used for the preparation of various baits formulation with the concept that cockroaches like to eat that food [66].

B. germanica preferred 1:3 protein-carbohydrate ratio food and self-selected optimal diets based upon their nutritional needs [69]. Ko et al. [15] determined that baits were the most effective when they matched the intake target and were preceded by food that departed from the intake target. When cockroaches were fed high-quality food and offered bait, they found the bait to be ineffective. It was determined that female German cockroaches compensated for their low dietary protein levels by increasing their consumption rates [55].

B. germanica had hygroreceptors in their antennae for detecting water vapor; however, the antennae were not capable of detecting odors or water vapor over more than a few centimeters [70]. German cockroaches are captured in different places depending on the availability of food or bait [71]. Rust et al. [13] did experiment by using sex pheromones to food-based odorants in German cockroaches. They determined that the direction to attractive baits was directed by olfaction over some distance using an olfactometer.

Silverman and Bieman [72] determined that the German cockroaches collected in different fields of Florida and South Korea showed avoidance behavior to bait formulation containing glucose, and substitution of glucose with D-fructose and/or a mixture of fructose: glucose in molar ratio $\geq$ 9:1 stimulated feeding behavior. Pol et al. [73] mentioned that German cockroaches were attracted to beer semiochemicals, ethanol, which was the product of formerly living and active yeast. In the process of production of beer yeast, actively metabolize the sugar in mated barley powder that was attractive to German cockroaches. In the laboratory experiment, they found that three components comprising the dry malt extract, water, and brewer's yeast strongly attracted the German cockroach. Similarly, the food intake of German and American cockroaches was stimulated by many sugars [74]. The sensilla on the maxillary palps of the American cockroach contained sugar-binding sites. Sugars have been frequently incorporated into insecticide-dosed diets as phagostimulants for cockroaches [75].

Frings [76] identified that in the American cockroach, the tarsi and cerci, lack gustatory receptors. The gustatory organs of cockroaches have both maxillary and labial palps. The maxillary palps play a major role in feeding. Moreover, he determined that American cockroaches responded behaviorally to several salts and acids.

Adler [77] reported that a bait made from the distilled grain was an attraction for the brown-banded cockroach but failed to attract American cockroaches during field experiments. Later Brenner and Patterson [78] tested the feeding preference of four species of cockroaches (American, smoky brown, brown-banded, and Florida wood cockroach) and found that brown-banded cockroach showed significant preference for cat chow and the brown-banded cockroach did not prefer distiller's grains when a choice was provided with other baits. Female brown-banded cockroaches' consumption rate was decreased when a 5% protein diet was provided, as compared to either 25% protein or commercial rat food [50].

9. Food stimulants

Phytophagous insects have evolved various mechanisms for detecting and avoiding consumption of certain allelochemicals [79]. Tsuji [80] assayed some fatty acid, their esters, and related alcohol in *P. americana*, *B. germamica,* and *Phragmatobia fuliginosa* and found that starved German cockroaches were attracted to n-caproic, n-caprylic, n-capric, and lauric saturated fatty acid, and oleic acid an unsaturated fatty acid. He further mentioned that the most effective attractants in esters were methyl myristate, ethyl myristate, methyl palmitate, and ethyl palmitate. Furthermore, he found that the saturated normal alcohols with 8 to 14 carbon atoms were also attractive for the American cockroach and oleyl alcohol was attractive to the German cockroach [80]. Moreover, the most effective feeding stimulants were n-caprylic acid and methyl myristate. Oleyl alcohol was the stimulant for German cockroaches only. Similarly, Wileyto and Boush [81] found that German cockroaches responded positively to oleyl alcohol, palmitic acid, fenugreek seed alcohol extract, and elaidic acid methyl ester in two choice olfactometer.

The attraction of German cockroaches to esters was tested, and it was found that hexyl hexanoate and pentanoate are attractive compounds and attractiveness was increased ten times with propyl cyclohexane acetate [82]. Moreover, Silverman and Selbach [19] determined that the glucose-averse strain of *B. germanica* rejected glucose solutions during brief exposure periods (< 5 min) for 2 days with food deprivation. Karimifar et al. [83] mentioned that 1-hexanol from peanut butter and ethanol and 2, 3-dihydro-3, 5-dihydroxy-6-methyl-4H-pyran-4-one (DDMP) from beer, are the key semiochemicals for making the food more attractive to *B. germanica.* It was determined that German cockroaches have hygroreceptors for detecting water vapor [70]; however, they were not detecting odors or water vapor or odors far from a few centimeters. Thus, the success of baiting appeared when located baits were meets maximum during the food searching period of the German cockroach [71].

Tsuji [74] studied the feeding behavior of three cockroach species to the constituents of rice bran and determined that volatile substances soluble in n-hexane acted as olfactory attractants. The feeding stimulants were methanol and the feeding stimulants were sugars and related compounds. The effects for galactose and mannitol were highly species-specific in *B. germanica* and *P. americana*. Jakinovich et al. [84] explained that the American cockroach had behavioral taste responses to sucrose and α-D-methyl-glucosides. In another experiment, Wieczorek [85] found that the sensilla field of the maxillary palps in the American cockroach has 2500 taste hairs. Insoluble α-glucoside was detected in the sensilla field of intact palps and hydrolyzed sugars were effective stimuli.

Sugarman and Jakinovich [86] did research on the behavioral gustatory response of the adult American cockroach and identified that the American cockroach responded positively to L-amino acids but not to most D-amino acids. Poultry liver, silkworm pupae, and hydrogenated soy protein were used as protein sources in bait matrices as feeding stimulants [87]. Tsuji [74] determined that attractiveness alone could not effective, it is more effective if it has both attractiveness and feeding stimulants as well which was found in n-hexane the soluble fraction of rice bran.

Moreover, it was found that certain attractants and feeding stimulants are species-specific and a mixture of these substances may have a synergistic effect [74, 80, 88]. Furthermore, [80] found that several fatty acids and related compounds were both attractive and served as feeding stimulants in *B. germanica, P. americana,* and *P. fuliginosa.* A mixture of glucose and fructose stimulated a higher feeding rate and a greater response of sugar gustatory receptor neurons in wild-type *B. germanica* than either fructose or glucose alone [88].

Nojima et al. [89] determined that the eighth tergal gland of male German cockroach secrets methanolic during sequential courtship behavior that contains seven compounds, including oligosaccharides mixture and maltose, maltotriose, and maltotetraose, that strongly stimulated the feeding response of females. Likewise, and an artificial blend of the sugar components significantly increased the polar lipid fraction in the gland and increased feeding stimulants in female cockroaches. Cohen et al. [45] and Ko et al. [15] also reported that higher protein, carbohydrate ratios could attract more German cockroaches. Furthermore, Ko et al. [15] used casein, peptone, and albumin as protein sources in their baits, which were more attractive to German cockroaches. Cohen et al. [45] used casein as the protein and determined that *B. germanica* and *S. longipalpa* preferred a higher ratio of casein to carbohydrates. Prakash et al. [90] identified that at 0.5 mg/cm^2 N, N-diethylphenyl acetamide exhibited residual repellency for 4, 3, and 2 weeks against American, German, and brown-banded cockroaches, respectively.

10. Baits

Ross [91] mentioned that the substance in the formation of bait has the most important role to determine attractiveness, whereas the inactive ingredient has a key role to determine the behavioral response in German cockroaches. For nymph 1–nymph 3 of the German cockroach, bred was found significantly more attractive than goliath gel, however, both were equally attractive for nymph 6 and adult females [17]. They further mention that there were no changes in the level of attractiveness of goliath gel bait either adding or removing fipronil in goliath gel baits. Therefore, several food types, mainly containing three macronutrients (carbohydrate, lipid, and protein) have been used in the food matrices of cockroach baits [92]. The feeding stimulant activity of fructose was less effective than glucose, sucrose, and maltose [20].

Durier and Rivault [17, 93] revealed that fipronil gel was more attractive than boric acid baits. Nalyanya et al. [16] mentioned that avert powder (abamectin), maxforce station and gel, and siege gel (all hydramethylnon) were regularly attractive to German cockroaches and brown-banded cockroach adults and nymphs in trapping experiments. Kaakeh et al. [94] stated that American cockroaches were more attracted to fipronil than combat bait matrix or to other alternative foods. Later, it was determined that German cockroaches frequently choose goliath gel bait (0.05% fipronil) in comparison to avert (0.05% abamectin), maxforce (2.15% hydramethylnon), and drax (33.3% boric acid) gel, and goliath gel induced more feeding stimulant than avert gels [17]. German cockroaches were effectively attracted to noviflumuron and fipronil baits; however, consumption was significantly higher in noviflumuron bait than fipronil baits under laboratory conditions [11].

Anaclerio and Molinari [95] investigated the attraction behavior of four synanthropic cockroach species of *B. germanica, S. longipalpa, Blatta orientalis,* and *P. americana*. Cockroach methanol fecal extracts showed a higher intraspecific attraction than aqueous extracts in olfactometer bioassays. They further mentioned that the new gel containing cockroach fecal extracts was more attractive than commercial gel formulations.

It was reported that German cockroaches were attractive and susceptible to fipronil and imidacloprid gel baits. At field levels, the German cockroach population was killed by 100% at 60 days after treatment with 0.05% fipronil and 2.15% imidacloprid gel baits [14, 96, 97].

11. Conclusions and future recommendations

This chapter describes the food detection and feeding behavior of three species of cockroaches (German, American, and brown-banded). The discussion was done on the sensitivity of the compound eyes of these species, specific food preferences of these species, as well as the efficacy of various baits in controlling these species.

It was found that the German cockroaches exhibited a bimodal activity response when they were exposed to natural light conditions, and the activity began at sunset and reached a peak several hours after the onset of darkness. Moreover, it was also found that a second peak activity occurred shortly before the light period began and was more noticeable under natural light than under artificial light. American cockroaches had diurnal locomotory activity correlated with alternating light and dark period; however, there was no correlation between temperature and humidity fluctuation. The brown-banded cockroach, calling occurred discontinuously and followed a diel pattern primarily during the scotophase, in a 12 L:12 h D photoperiod and after transfer to continuous light or dark conditions.

The German cockroach can balance the micronutrient intake by selecting a mixed diet from foods. American cockroaches were able to associate food odors with proteins. Moreover, it was determined that this species exhibited a spontaneous initial preference for vanilla over menthol. Likewise, brown-banded cockroach self-selected a 15.5:84.5 protein: carbohydrate diet.

The German cockroach males foraged for food or water at least once in 7 days out of 10 days, whereas gravid females foraged actively on fewer than 3 days. With their fat and water reserves, female German cockroaches are well-suited to survive without food on a temporary basis. The possession of specific hygroreceptors could play an important role in the food searching behavior of the American cockroach. In the absence of drinking water, the nymphs of brown-banded cockroaches survived significantly longer than German cockroach nymphs.

Cockroach feeding was affected by sex, age, and reproductive condition. The feeding behaviors of female *B. germanica* differed from nymph and adult males. When resources were farther away from the shelter the number of drinking and eating bouts increased. It was higher and took longer than at lower German cockroach population densities. Male German cockroaches were more active and explored more than females. Feeding and drinking activity peaked during the egg maturation period and ended sharply at the appearance of egg capsules.

German cockroaches had the color vision and the dorsal part of the compound eye had two peak sensitivities. The main visual wavelength spectrum was in the blue-green portion of the spectrum (490 nm), while the minor one was in the ultraviolet. The compound eyes of *P. americana* contained at least two kinds of receptors. They were stimulated by the blue-green and violet (or ultraviolet) regions of the spectrum. The dark-adapted American cockroach ocellus was dependent on the intensity, but not on the wavelength of stimulating light.

Water was required more critically than food for German cockroaches. Male *B. germanica* could not survive more than 9–10 days, but females could survive with access to water and without food for up to 45 days.

Food also played an important role in the temporal and spatial distribution of population densities of the German cockroach. German cockroach baits are a mixture of odor or pheromone lures that attract the insect. Two percent Dipterex bait and powdered sugar showed the greatest attractiveness toward German cockroaches. Dextrin and corn starch alone or in combination were the most attractive to nymphs of American cockroaches in laboratory condition. The food intake of German and American cockroaches was stimulated by many sugars. Bait made from the distilled grain was attractive for the brown-banded cockroach but failed to attract American cockroaches during field experiments.

B. germanica had hygroreceptors in their antennae for detecting water vapor; however, the antennae were not capable of detecting odors or water vapor over more than a few centimeters. The sensilla on the maxillary palps of the American cockroach contained sugar-binding sites. It was found that starved German cockroaches were attracted to n-caproic, n-caprylic, n-capric, and lauric saturated fatty acid, and oleic acid an unsaturated fatty acid. He further mentioned that the most effective attractants in esters were methyl myristate, ethyl myristate, methyl palmitate, and ethyl palmitate. Furthermore, it was found that the saturated normal alcohols with 8 to 14 carbon atoms were also attractive for the American cockroach and oleyl alcohol was attractive to the German cockroach. The effects of galactose and mannitol were highly species-specific in the German and American cockroaches. The several fatty acids and related compounds were both attractive and served as feeding stimulants in German and American cockroaches.

Avert powder (abamectin), maxforce station and gel, and siege gel (all hydramethylnon) were regularly attractive to German cockroach and brown-banded cockroach adults and nymphs in the trapping experiment. American cockroaches were more attracted to fipronil than combat bait matrix.

For the control of cockroaches and developing modern baits, specific knowledge on circadian rhythm, food searching behavior, eating, and drinking preferences, effect of population density on feeding and drinking behavior, factors affecting foraging and feeding behavior, visual sensitivity, food attractants, and stimulants should be considered. It is clear that very limited studies were carried out on the sensitivity of the compound eyes and food detection ability and the feeding behavior of brown-banded cockroaches. Thus, further research is needed in those areas for the control of brown-banded cockroaches.

Author details

Anil Chandra Neupane
Adventist Development and Relief Agency in Nepal (ADRA), Lalitpur, Nepal

*Address all correspondence to: anilnp111@gmail.com

References

[1] Gore JC, Schal C. Cockroach allergen biology and mitigation in the indoor environment. Annual Review of Entomology. 2007;**52**:439-463

[2] Sarinho E, Schor D, Veloso M, Rizzo J. There are more asthmatics in homes with high cockroach infestation. Brazilian Journal of Medical and Biological Research. 2004;**37**:503-510

[3] Wu C, Lee M, Wang NM. Expression of the American cockroach per a 1 allergen in mammalian cells. Allergy. 2000;**55**:1179-1183

[4] Schal C, Hamilton R. Integrated suppression of synanthropic cockroaches. Annual Review of Entomology. 1990;**35**:521-551

[5] Fakoorziba M, Eghbal F, Hassanzadeh J, Moemenbellah-Fard M. Cockroaches (Periplaneta americana and Blattella germanica) as potential vectors of the pathogenic bacteria found in nosocomial infections. Annals of Tropical Medicine & Parasitology. 2010;**104**:521-528

[6] Fathpour H, Emtiazi G, Ghasemi E. Cockroaches as reservoirs and vectors of drug resistant salmonella spp. Iranian Biomedical Journal. 2003;**7**:35-38

[7] Fotedar R, Shriniwas UB, Verma A. Cockroaches (Blattella germanica) as carriers of microorganisms of medical importance in hospitals. Epidemiology & Infection. 1991;**107**:181-187

[8] Nasirian H. New aspects about Supella longipalpa (Blattaria: Blattellidae). Asian Pacific Journal of Tropical Biomedicine. 2016;**6**:1065-1075

[9] Pai H-H, Chen W-C, Peng C-F. Isolation of bacteria with antibiotic resistance from household cockroaches (Periplaneta americana and Blattella germanica). Acta Tropica. 2005a;**93**:259-265

[10] Vazirianzadeh B, Dehghani R, Mehdinejad M, Sharififard M, Nasirabadi N. The first report of drug resistant bacteria isolated from the brown-banded cockroach, Supella longipalpa, in Ahvaz, South-Western Iran. Journal of Arthropod-Borne Diseases. 2014;**8**:53

[11] Wang C, Bennett GW. Efficacy of noviflumuron gel bait for control of the German cockroach, Blattella germanica (Dictyoptera: Blattellidae)—Laboratory studies. Pest Management Science. 2006b;**62**:434-439

[12] Appel AG. Laboratory and field performance of consumer bait products for German cockroach (Dictyoptera: Blattellidae) control. Journal of Economic Entomology. 1990;**83**:135-159

[13] Rust MK, Owens JM, Reierson DA. Understanding and Controlling the German Cockroach. New York: Oxford University Press, Inc; 1995

[14] Appel A, Tanley M. Laboratory and field performance of an imidacloprid gel bait against German cockroaches (Dictyoptera: Blattellidae). Journal of Economic Entomology. 2000;**93**:112-118

[15] Ko AE, Schal C, Silverman J. Diet quality affects bait performance in German cockroaches (Dictyoptera: Blattellidae). Pest Management Science. 2016;**72**:1826-1836

[16] Nalyanya G, Liang D, Kopanic RJ, Schal C. Attractiveness of insecticide baits for cockroach control (Dictyoptera:

Blattellidae) laboratory and field studies. Journal of Economic Entomology. 2001;**94**:686-693

[17] Durier V, Rivault C. Comparisons of toxic baits for controlling the cockroach, Blattella germanica: Attractiveness and feeding stimulation. Medical and Veterinary Entomology. 2000;**14**:410-418

[18] Silverman J, Ross MH. Behavioral resistance of field-collected German cockroaches (Blattodea: Blattellidae) to baits containing glucose. Environmental Entomology. 1994;**23**:425-430

[19] Silverman J, Selbach H. Feeding behavior and survival of glucose-averse Blattella germanica (Orthoptera: Blattoidea: Blattellidae) provided glucose as a sole food source. Journal of Insect Behavior. 1998;**11**:93-102

[20] Wang C, Scharf ME, Bennett GW. Behavioral and physiological resistance of the German cockroach to gel baits (Blattodea: Blattellidae). Journal of Economic Entomology. 2004;**97**:2067-2072

[21] Wang C, Scharf ME, Bennett GW. Genetic basis for resistance to gel baits, fipronil, and sugar-based attractants in German cockroaches (Dictyoptera: Blattellidae). Journal of Economic Entomology. 2006;**99**:1761-1767

[22] Milio J, Koehler P, Patterson R. Laboratory and field evaluations of hydramethylnon bait formulations for control of American and German cockroaches (Orthoptera: Blattellidae). Journal of Economic Entomology. 1986;**79**:1280-1286

[23] Holbrook GL, Roebuck J, Moore CB, Waldvogel MG, Schal C. Origin and extent of resistance to fipronil in the German cockroach, Blattella germanica (L.) (Dictyoptera: Blattellidae). Journal of Economic Entomology. 2003;**96**:1548-1558

[24] Liang D. Year. PublishedPerformance of Cockroach Gel Baits Against Susceptible and Bait averse Strain of German Cockroach, *Blatella germanic* (Dictyoptera: Blattellidae)-Role of Bait Base and Active Ingredient. International conference on urban pests; 2005

[25] Pai H-H, Wu S-C, Hsu E-L. Insecticide resistance in German cockroaches (Blattella germanica) from hospitals and households in Taiwan. International Journal of Environmental Health Research. 2005b;**15**:33-40

[26] Agrawal VK, Agarwal A, Choudhary V, Singh A, Agrawal P. Efficacy of fipronil gel over imlprothrin+ cypermethrin aerosol in control of cockroaches (field trial). Annals of Tropical Medicine and Public Health. 2012;**5**:423

[27] Fardisi M, Gondhalekar AD, Scharf ME. Development of diagnostic insecticide concentrations and assessment of insecticide susceptibility in German cockroach (Dictyoptera: Blattellidae) field strains collected from public housing. Journal of Economic Entomology. 2017;**110**:1210-1217

[28] Liang D, McGill J, Pietri JE. Unidirectional cross-resistance in German cockroach (Blattodea: Blattellidae) populations under exposure to insecticidal baits. Journal of Economic Entomology. 2017;**110**:1713-1718

[29] Raubenheimer D, Jones S. Nutritional imbalance in an extreme generalist omnivore: Tolerance and recovery through complementary food selection. Animal Behaviour. 2006;**71**:1253-1262

[30] Tsai CW, Lee HJ. Circadian locomotor rhythm masked by the female

reproduction cycle in cockroaches. Physiological Entomology. 2000;**25**:63-73

[31] Metzer R. Behavior. In: Rust MK, Owens JM, Reierson DA, editors. Understanding and Controlling the German Cockroach. New York: Oxford University Press, Inc.; 1995. pp. 49-76

[32] Lauprasert P, Sitthicharoenchai D, Thirakhupt K, Pradatsudarasar A-O. Food preference and feeding behavior of the German cockroach, Blattella germanica (Linnaeus). Journal Science Research of Chulalongkorn University. 2006;**31**:121-126

[33] Fuchs M. Behaviour analysis and development of baiting methods at populations of Blattella germanica and Blatta orientalis. Praktische Schaedlingsbekaempfer (Germany, FR). 1983. ISSN: 0032-6801

[34] Sommer S. Experimentelle Untersuchungen zur circadianen lokomotorischen Aktivitat von Blattella germanica L.(Dictyopt., Blattellidae). Biologisches Zentralblatt. 1978;**973**:337-343

[35] Silverman J. Adult German cockroach (Orthoptera: Blattellidae) feeding and drinking behavior as a function of density and harborage-to-resource distance. Environmental Entomology. 1986;**15**:198-204

[36] Rivault C, Cloarec A. Exploitation of food resources by the cockroach Blattella germanica in an urban habitat. Entomologia Experimentalis et Applicata. 1991;**61**:149-158

[37] Nalyanya G, Moore CB, Schal C. Integration of repellents, attractants, and insecticides in a" push-pull" strategy for managing German cockroach (Dictyoptera: Blattellidae) populations. Journal of Medical Entomology. 2000;**37**:427-434

[38] Lipton G, Sutherland D. Activity rhythms in the American cockroach, Periplaneta americana. Journal of Insect Physiology. 1970;**16**:1555-1566

[39] Cloudsley-Thompson J. LXIX.—Studies in diurnal rhythms.—III. Photoperiodism in the cockroach Periplaneta americana (L.). Journal of Natural History. 1953;**6**:705-712

[40] Harker JE. Factors controlling the diurnal rhythm of activity of Periplaneta americana L. Journal of Experimental Biology. 1956;**33**:224-234

[41] Smith AF, Schal C. Circadian calling behavior of the adult female brown-banded cockroach, Supella longipalpa (F.)(Dictyoptera: Blattellidae). Journal of Insect Behavior. 1991;**4**:1-14

[42] Ebeling W, Wagner R, Reierson DA. Influence of repellency on the efficacy of blatticides. I. Learned modification of behavior of the German cockroach. Journal of Economic Entomology. 1966;**59**:1374-1388

[43] Gadd CA, Raubenheimer D. Nutrient-specific learning in an omnivorous insect: The American cockroach Periplaneta americana L. learns to associate dietary protein with the odors citral and carvone. Journal of Insect Behavior. 2000;**13**:851-864

[44] Balderrama N. One trial learning in the American cockroach, Periplaneta americana. Journal of Insect Physiology. 1980;**26**:499-504

[45] Cohen R, Heydon S, Waldbauer G, Friedman S. Nutrient self-selection by the omnivorous cockroach Supella longipalpa. Journal of Insect Physiology. 1987;**33**:77-82

[46] Kunkel JG. Development and the availability of food in the German

cockroach, Blattella germanica (L.). Journal of Insect Physiology. 1966;**12**:227-235

[47] Rivault C. Spatial distribution of the cockroach, Blattella germanica, in a swimming-bath facility. Entomologia Experimentalis et Applicata. 1989;**53**:247-255

[48] Durbin EJ, Cochran DG. Food and water deprivation effects on reproduction in female Blattella germanica. Entomologia Experimentalis et Applicata. 1985;**37**:77-82

[49] Melton RH. Differential adaptation to water deprivation in first-instar nymphs of the German cockroach (Blattella germanica) and the brown-banded cockroach (Supella longipalpa). Entomologia Experimentalis et Applicata. 1995;**77**:61-68

[50] Hamilton RL, Cooper RA, Schal C. The influence of nymphal and adult dietary protein on food intake and reproduction in female brown-banded cockroaches. Entomologia Experimentalis et Applicata. 1990;**55**:23-31

[51] Cloarec A, Rivault C. Age-related changes in foraging in the German cockroach (Dictyoptera: Blattellidae). Journal of Insect Behavior. 1991;**4**:661-673

[52] Ballard J, Ball H, Gold R. Influence of selected environmental factors upon German cockroach (Orthoptera: Blattellidae) exploratory behavior in choice boxes. Journal of Economic Entomology. 1984;**77**:1206-1210

[53] Cochran DG. Food and water consumption during the reproductive cycle of female German cockroaches. Entomologia Experimentalis et Applicata. 1983;**34**:51-57

[54] Keil CB. Structure and estimation of shipboard German cockroach (Blattella germanica) populations. Environmental Entomology. 1981;**10**:534-542

[55] Hamilton RL, Schal C. Effects of dietary protein levels on reproduction and food consumption in the German cockroach (Dictyoptera: Blattellidae). Annals of the Entomological Society of America. 1988;**81**:969-976

[56] Koehler P, Agee H, Leppla N, Patterson R. Spectral sensitivity and behavioral response to light quality in the German cockroach (Dictyoptera: Blattellidae). Annals of the Entomological Society of America. 1987;**80**:820-822

[57] Guthrie D, Tindall A. The Biology of the Cockroach. New York: St. Martin's; 1968

[58] Walther JB. Changes induced in spectral sensitivity and form of retinal action potential of the cockroach eye by selective adaptation. Journal of Insect Physiology. 1958;**2**:142-151

[59] Mote MI, Goldsmith TH. Spectral sensitivities of color receptors in the compound eye of the cockroach Periplaneta. Journal of Experimental Zoology Part A: Ecological Genetics and Physiology. 1970;**173**:137-145

[60] Goldsmith TH, Ruck PR. The spectral sensitivities of the dorsal ocelli of cockroaches and honeybees: An electrophysiological study. The Journal of General Physiology. 1958;**41**:1171-1185

[61] Willis ER, Lewis N. The longevity of starved cockroaches. Journal of Economic Entomology. 1957;**50**:438-440

[62] Karimifar N. Semiochemical-Based Food-Foraging in German Cockroaches, Blattella germanica L.(Dictyoptera: Blattellidae). Dept. of Biological Sciences-Simon Fraser University. 2009;**89**

[63] Barcay S, Schneider B, Bennett G. Influence of Insecticide treatment on German cockroach (Dictyoptera: Blattellidae) movement and dispersal within apartments. Journal of Economic Entomology. 1990;**83**:142-147

[64] Owens JM, Bennett GW. Comparative study of German cockroach (Dictyoptera: Blattellidae) population sampling techniques. Environmental Entomology. 1983;**12**:1040-1046

[65] Wang C, Bennett GW. Comparison of cockroach traps and attractants for monitoring German cockroaches (Dictyoptera: Blattellidae). Environmental Entomology. 2006a;**35**:765-770

[66] Reierson A. Baits for German cockroach control. Understanding and controlling the German cockroach. 1995:231-265

[67] Ballard JB, Gold RE. The effect of selected baits on the efficacy of a sticky trap in the evaluation of German cockroach populations. Journal of the Kansas Entomological Society. 1982;**55**:86-90

[68] Lofgren C, Burden GS. Tests with poison baits against cockroaches. The Florida Entomologist. 1958;**41**:103-110

[69] Shik JZ, Schal C, Silverman J. Diet specialization in an extreme omnivore: Nutritional regulation in glucose-averse German cockroaches. Journal of Evolutionary Biology. 2014;**27**:2096-2105

[70] Altner H, Loftus R. Ultrastructure and function of insect thermo-and hygroreceptors. Annual Review of Entomology. 1985;**30**:273-295

[71] Ebeling W, Reierson D. Effect of population density on exploratory activity and mortality rate of German cockroaches in choice boxes. Journal of Economic Entomology. 1970;**63**:350-355

[72] Silverman J, Bieman D. Issues affecting the performance of cockroach baits. In: Proc 2nd Internat Conf on Insect Pests in the Urban Environment, Heriot-Watt University, Edinburgh, Scotland. Citeseer; 1996. pp. 341-346

[73] Pol JC, Jimenez SI, Gries G. New food baits for trapping German cockroaches, Blattella germanica (L.) (Dictyoptera: Blattellidae). Journal of Economic Entomology. 2017;**110**:2518-2526

[74] Tsuji H. Studies on the behaviour pattern of feeding of three species of cockroaches, Blattella germanica (L.), Periplaneta americana L., and P. fuliginosa S., with special reference to their responses to some constituents of rice bran and some carbohydrates. Medical Entomology and Zoology. 1965;**16**:255-262

[75] Becker A, Peters W. Localization of sugar-binding sites in contact chemosensilla of Periplaneta americana. Journal of Insect Physiology. 1989;**35**:239-250

[76] Frings H. Gustatory thresholds for sucrose and electrolytes for the cockroach, Periplaneta americana (Linn.). Journal of Experimental Zoology Part A: Ecological Genetics and Physiology. 1946;**102**:23-50

[77] Adler VE. A highly effective attractant for the brownbanded cockroach (Orthoptera: Blattellidae). Journal of Environmental Science & Health Part A. 1985;**20**:839-844

[78] Brenner RJ, Patterson RS. Laboratory feeding activity and bait preferences of four species of cockroaches (Orthoptera: Blattaria). Journal of Economic Entomology. 1989;**82**:159-162

[79] Schoonhoven L, Meerman J. Metabolic cost of changes in diet and neutralization of allelochemics.

Entomologia Experimentalis et Applicata. 1978;**24**:689-693

[80] Tsuji H. Attractive and feeding stimulative effect of some fatty acids and related compounds on three species of cockroaches. Medical Entomology and Zoology. 1966;**17**:89-97

[81] Wileyto PE, Boush MG. Attraction of the German cockroach, Blattella germanica (Orthoptera: Blatellidae), to some volatile food components. Journal of Economic Entomology. 1983;**76**:752-756

[82] Sugawara R, Kurihara S, Muto T. Attraction of the German cockroach to cyclohexyl alkanoates and n-alkyl cyclohexaneacetates. Journal of Insect Physiology. 1975;**21**:957-964

[83] Karimifar N, Gries R, Khaskin G, Gries G. General food semiochemicals attract omnivorous German cockroaches, Blattella germanica. Journal of Agricultural and Food Chemistry. 2011;**59**:1330-1337

[84] Jakinovich W, Sugarman D, Vlahopoulos V. Gustatory responses of the cockroach, house fly, and gerbils to methyl glycosides. Journal of Comparative Physiology. 1981;**141**:297-301

[85] Wieczorek H. Biochemical and behavioral studies of sugar reception in the cockroach. Journal of Comparative Physiology. 1978;**124**:353-356

[86] Sugarman D, Jakinovich JW. Behavioural gustatory responses of adult cockroaches, Periplaneta americana to D and L amino acids. Journal of Insect Physiology. 1986;**32**:35-41

[87] Wolfe J, Lesiewicz D, Mehra Y, Mares J. Cockroach Bait Feeding Stimuli. Google Patents; 1997. Retrieved on Dec. 16, 2019, from the World Wide Web: https://patents.google.com/patent/US5676961A/en

[88] Wada-Katsumata A, Silverman J, Schal C. Changes in taste neurons support the emergence of an adaptive behavior in cockroaches. Science. 2013;**340**:972-975

[89] Nojima S, Nishida R, Kuwahara Y, Sakuma M. Nuptial feeding stimulants: A male courtship pheromone of the German cockroach, Blattella germanica (L.)(Dictyoptera: Blattellidae). Naturwissenschaften. 1999;**86**:193-196

[90] Prakash S, Srivastava C, Kumar S, Pandey K, Kaushik M, Rao K. N, N-diethylphenylacetamide—A new repellent for Periplaneta americana (Dictyoptera: Blattidae), Blattella germanica, and Supella longipalpa (Dictyoptera: Blattellidae). Journal of Medical Entomology. 1990;**27**:962-967

[91] Ross MH. Laboratory studies on the response of German cockroaches (Dictyoptera: Blattellidae) to an abamectin gel bait. Journal of Economic Entomology. 1993;**86**:767-771

[92] Tee H, Lee C. Sustainable cockroach management using insecticidal baits: Formulations, behavioral responses and issues. Urban Insect Pests-Sustainable Management Strategies; Dhang, P., Ed. 2014;**1**:65-85

[93] Durier V, Rivault C. Effect of spatial knowledge and feeding experience on foraging choices in German cockroaches. Animal Behaviour. 2001;**62**:681-688

[94] Kaakeh W, Reid BL, Bennett GW. Toxicity of fipronil to German and American cockroaches. Entomologia Experimentalis et Applicata. 1997;**84**:229-237

[95] Anaclerio M, Molinari F. Intra and inter-specific attraction of cockroach faecal extracts: Studies for improving bait activity. Bulletin of Insectology. 2012;**65**:113-118

[96] Agrawal V, Tilak R. Field performance of imidacloprid gel bait against German cockroaches (Dictyoptera: Blatellidae). Indian Journal of Medical Research. 2006;**124**:89

[97] Nasirian H. Duration of fipronil and imidacloprid gel baits toxicity against Blattella germanica strains of Iran. Journal of Arthropod-Borne Diseases. 2007;**1**:40-47

Chapter 4

Interaction Among the Multi-Trophic Lac Insect Complex of Flora and Fauna: Impact on Quantity and Quality of the Resin Secreted

Kewal Krishan Sharma and Thamilarasi Kandasamy

Abstract

Lac insects are a specialized group of phytosuccivorous insects (Coccoidea: Tachardiidae) that secret resin of industrial importance having diverse applications. Due to unique biology, host preference and dispersal mechanisms, lac insects are expected to differentiate locally, forming geographic and host races without adequate morphological differentiation. 101 species of lac-insects and over 400 species of lac host plants have been reported but insects belonging to sub-family Tachardiinae are considered important for *laksha-culture* (lac insect farming). With a wide host-plant range and diverse habitat, the insects have developed a specialized ecosystem with multi-trophic complex of flora and fauna. Not only the lac insect but also the host plants and various biotic associations play a significant role in determining the quantity and quality of the produce. This insect being an obligate phloem sap sucker completes its life cycle on host plant species. Phloem sap is nutritionally unbalanced, as it is rich in carbohydrates but deficient in essential amino acids. Due to the scarcity of essential elements in phloem sap, endosymbionts are likely to co-evolve within the insect cell, while fulfilling their nutritional requirement. Implication of these intricate biotic associations on quantity and quality of the lac resin produced merits thorough understanding for sustained lac production.

Keywords: lac insect, *Kerria lacca*, Tachardiidae, insect-plant interaction, resin quantity and quality, lac-endosymbionts

1. Introduction

Lac, reputed as the only resin of animal origin is secreted mainly by the Indian lac insect, *Kerria lacca* (Kerr) (Hemiptera: Tachardiidae), which thrives on the tender twigs of specific trees called lac hosts (*kusum*, *palas*, *ber*, *Flemingia etc.*). Since time immemorial, lac farming has been practised for the products of commerce *viz.*, lac, a resinous or non-resinous covering substance over their body, dye – a natural crimson/

yellow color in the body fluid and the lac wax – present within and above the lac resin. These products find application in diverse areas such as food, pharmaceuticals, cosmetics, paints and varnish industries [1]. Lac insects (Hemiptera: Coccomorpha) specialized coccids (scale insects) belonging to the family Tachardiidae (=Kerriidae) [2, 3] are sap-sucking insects thriving on certain plant species called lac host plants. Almost all the stages of lac insect are sedentary and attached to the host plants except for neonate nymphs (crawlers) and adult male insects. Lac insects can be found in tropical and subtropical regions (between the latitudes 40^0 N and 40^0 S) due to their preference for a warm climate. Lac production is mainly done in some South, East and Southeast Asian countries like India, Thailand, China, Indonesia, Bangladesh, Myanmar, Laos and Vietnam but the product is in demand all over the world. India is the global leader in lac production followed by Thailand.

The family comprises of nine genera and 99 species worldwide [4]; *Kerria* Targioni Tozzetti is the largest genus in the family. Recently two more species *K. destructor* [5] and *K. canalis* [6] have been described taking the total number of species to 101. The family is characterized by sclerotized features of the adult female. The outer lac encrustation, called lac cell, does not always help in recognition, although it provides indicators in some cases, *e.g.*, lac is resinous and alcohol soluble in sub-family Tachardiinae, but hard, horny and insoluble in Tachardiininae.

India is endowed with a rich wealth of lac insect resources. The genus *Kerria* includes 28 species worldwide, 21 of which are recorded from India [7]. 27.7% of lac insect biodiversity reported from the world is found in our country under two genera *i.e.*, *Kerria* (23 species) and *Paratachardina* (5 species). Species of *Paratachardina* do not produce true lac and are considered pests of economically important plants but have been utilized as bio-control agents for controlling weeds. Other minor species include *K. chinensis* and *K. sharda*. *K. chinensis* is the principal insect in East and Southeast Asian countries. The *kusmi* form of *Kerria lacca*, known for its superior quality of lac and higher productivity is unique to India.

Strains of lac insects: Indian lac insect is known to comprise two distinct infra subspecific forms (commonly termed strains), '*Kusmi*' and '*Rangeeni*' [8]. *Kusmi* strain is grown on *Kusum* or on other alternate host plants and the *kusmi* crops are (i) *Jethwi* – summer season (maturing in June/July) and (ii) *Aghani* – winter season (maturing in Jan./Feb.). *Rangeeni* strain thrives on host plants like *palas* but not on *kusum* and it has also two crops; they are (i) *Katki* – rainy season (maturing in Oct. /Nov.) and (ii) *Baisakhi* – summer season (maturing in June/July).

2. Factors affecting lac production

Production of lac per unit area and time depends on various biotic and abiotic factors impinging upon lac insect ecosystem. Important biotic factors, which affect lac productivity are:

2.1 Lac insects

Species mainly belonging to *Kerria* are exploited for commercial production of lac. Yield of lac insect varies significantly depending upon the lac insect species and its strain. Differences in lac yield also exist between the two strains. Average resin secreted by *kusmi* cells and *rangeeni* cells is 16.96 mg and 8.07 mg, respectively.

Similarly, *kusmi* strain produced twice the quantity of resin (4.99 mg/mm) per unit size of the cell compared to *rangeeni* strain (2.55 mg/mm) [9].

2.1.1 Initial density of settlement and mortality

Lac insects are gregarious in nature and settle in close proximity. Hence, density of settlement has a bearing on lac yield. Different lac insects showed a varied density of settlement (average 80−192 per sq. cm in *rangeeni,* 186−242 per sq. cm in *kusmi* and 160–264 per sq. cm in Meghalaya stock) based on the broodlac quantity used [10]. Under the optimal broodlac conditions, *kusmi* strain tends to settle closer compared to *rangeeni* strain, which is evident from their mean density of settlement. However, under excess brood condition 'crowding effect' is exhibited, which is more marked in *rangeeni.* On the other hand, crawlers of *K. chinensis* (Meghalaya collection) tend to settle closer during both the seasons. Settlement of larvae in variance with the desired number affects the production adversely. Higher density of settlement results either in increased mortality due to insufficient availability of space and nutrition or a higher male population, which ultimately affects lac yield. Mortality up to 21 days of inoculation is attributed to non-feeding of the larvae at the time of initial settlement. Higher initial mortality is indicative of non-suitability of the host plant and/or of unfavorable environmental conditions.

2.1.2 Sex-ratio

Although the contribution of male lac insects to commercial lac production is very little, they are vital for good lac crop and vis-à-vis broodlac production since the rate of lac secretion increases vigorously in the females after fertilization. However, female lac insects are the sole commercial lac producers. Progeny size and sex ratio vary widely in different lac insect stocks as well as when same lac insect is reared on different host plants. Variation in sex ratio in different crops of lac insect vis-à-vis host plant is more pronounced in *rangeeni* strain than in *kusmi,* especially during the summer season. It varied between 18.07–64.39 and 25.59–31.48%, respectively, for summer and rainy season crops of *rangeeni* and between 22.02–29.33 and 25.49–35.44%, respectively, for summer and winter season crops of *kusmi* strain [10].

Average male population is lower in smaller progenies and higher in larger progenies; suggesting that larger progeny size increases the male population. In general, sex ratio ranges between 20 and 50% depending on various biotic and abiotic factors. However, in smaller colonies, it may vary between 0 and 100 per cent. Sex ratio is found to vary with (i) season [11]; (ii) sequence of emergence [12]; (iii) site of colonization [12]; iv) density of settlement [13, 14]; v) plant-host [15] and plant-host variety [16]. However, the exact reasons for the wide fluctuation of sex ratio in lac insect population need to be investigated further.

2.2 Host-plant

Although more than 400 species of plants have been reported to support lac insects on them, various biological attributes such as survival, resin production and fecundity differ greatly from the host species used [17]. Coccoids have been found to be very specific not only to different host species but also to specific varieties and even individual phenotypes of host plants due to inter-specific and intra-specific variation in the host plant defense. Srinivasan [18] has indicated preference of lac insects to specific phenotypes – *kariya* over *charka* – the later has a lighter colored bark than the former in *palas* (*Butea monosperma*) and *Kusum* (*Schleichera oleosa*); however, the two are botanically inseparable.

Food quality: The quality and quantity of food available to the insect are important in determining its survival and reproduction rate. This is particularly true for phloem sap-feeding insects. Passive exudation of sap of phloem bundles has a substantial role in supply of phloem sap to the insects. Quantitative and qualitative differences in the nutrition available from different host plants, cause variation in biological attributes of the lac insect. Variability studies in case of four species of *Flemingia viz. macrophylla*, *semialata*, *stricta* and *bractiata* with regard to various attributes of lac insects have shown significant differences [19].

Sap condition: Lac insect feeds on the phloem sap of the host plant. The insect inserts its proboscis and feeds on exudation of sap by i) turgor pressure and (ii) capillary action. Turgor pressure of the host changes with the season and phenotypic activity of the plant. The sap pressure is higher during rainy season and considered favorable for the growth of the insect and inverse is true for summer season. 'Sap reaction' and 'sap density' are possibly among the factors, which influence the suitability of the host plant for lac infection. Good host plants have phloem sap of pH ranging between 5.8 and 6.2 [20]. Similarly, sap density of good hosts ranges between 0.14–0.1728.

2.2.1 Initial density of settlement and mortality

Initial settlement of crawlers is affected mainly by host plants and the physical characteristics of the twigs where they settle down. Density of settlement on *S. oleosa* (*kusum*) is higher than on *Albizia lucida* (*Galwang*). Some lac host plants support a certain species or strain in a better way and *vice-versa*. Although lac insect crawlers can be made to settle on any plant twigs, they would be able to survive and complete their life cycle only on good hosts. *Rangeeni* lac insect cannot survive on *kusum* and it exhibits very high mortality on *F. semialata*; similarly, *kusmi* strain cannot survive on *B. monosperma* (*palas*), while *Ziziphus mauritiana* (*ber*) supports both strain up to maturity. There is a significant decrease in survival when lac insect was reared on pumpkin fruits (*Cucurbita moschata*) in comparison to *Flemingia macrophylla*. The increase in mortality observed was 67.2% and 104.9% for *kusmi* and *rangeeni* strains, respectively [21].

2.2.2 Sex-ratio

Chauhan [15] has reported that the Meghalaya lac insect stock showed significant difference in sex ratio on different host plants. It was observed that 72%, 82% and 98% were in favor of males on *F. macrophyll*a, *Cajanus cajan* and *Z. mauritiana*, respectively. Similarly, Sharma and Ramani [21] have also observed that male percentage of *rangeeni* and *kusmi* strains of *K. lacca* on *F. macrophylla* was 39.76 and 37.28%, which increased to 70.05 and 62.65% when they were reared on *C. moschata* fruits.

2.2.3 Effect of host-plant on resin production

Average mean cell diameter of a *kusmi* female cell was 3.02, 3.16, 3.50 and 3.54 mm on *C. moschata*, *F. macrophylla, Acacia auriculiformis* and *S. oleosa,* respectively [9]. Resin produced by individual female lac cells varied significantly. It ranged from

6.11 mg on *C. moschata* fruits to 22.84 mg on *S. oleosa.* Resin production by individual female lac insects was the highest on *S. oleosa* followed by *A. auriculiformis, F. macrophylla* and *C. moschata* fruits (**Figure 1**). Very high intra-strain variations were observed in resin-producing efficiency of lac insect even when cultured on the same host plant.

Similarly, the average mean diameter of a *rangeeni* female cell grown on *A. auriculiformis, C. moschata, F. macrophylla* and *B. monosperma* was 3.10, 3.17, 3.19 and 3.22 mm, respectively. Average resin secreted by an individual *rangeeni* female cell ranged between 6.00 mg (on *C. moschata* fruits) and 9.09 mg (on *A. auriculiformis*). Resin production by single *rangeeni* female lac insect was found to be the highest on *A. auriculiformis* followed by *B. monosperma, F. macrophylla* and the lowest on *C. moschata* fruits (**Figure 2**). Higher values of coefficient of regression in good hosts *S. oleosa* and *A. auriculiformis* for *kusmi* strain and *B. monosperma* in *rangeeni* strain corroborate the fact that a good lac host allows full manifestation of the resin-producing potential of the lac insect. Resin productivity is higher on tree hosts in comparison to *F. macrophylla* (a shrub) and the pumpkin fruit. Though variations in cell size were less prominent, weight of the cell and resin output per female recorded greater variations showing the effect of host on resin productivity of the insect. Lac insect – host plant interaction in terms of lac production and host suitability is reflected in the data provided in **Table 1.**

2.2.4 Host suitability index

Host preference, length of settlement at crop maturity, lac insect survival at crop maturity and resin production by the insect are the most important attributes affecting the production of lac. Host Suitability Index is calculated for identifying an ideal host plant by using the following formula. By taking the lowest value of Host Suitability Index for a particular host plant as 1.00, relative suitability indices are calculated for the other hosts.

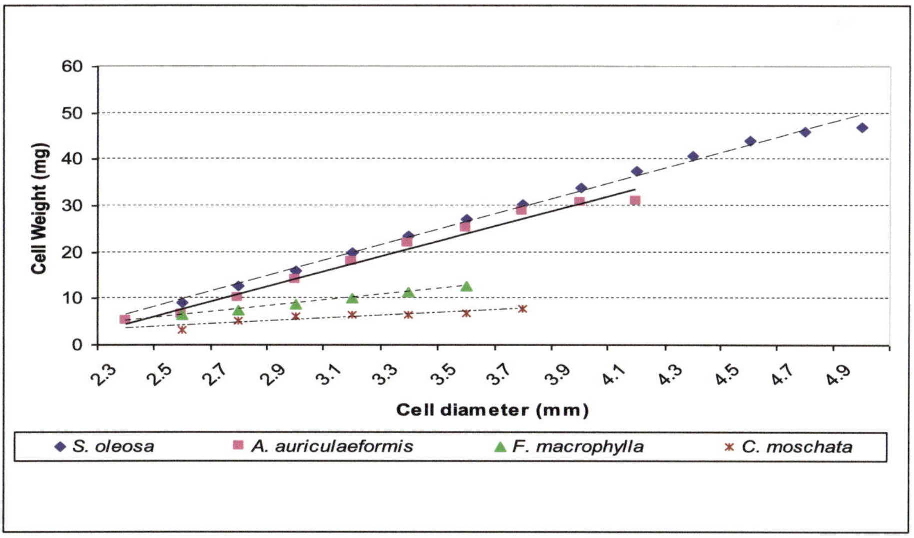

Figure 1.
Resin productivity and cell size relationship in Kerria lacca *(*kusmi *strain) on different host plants during winter season crop.*

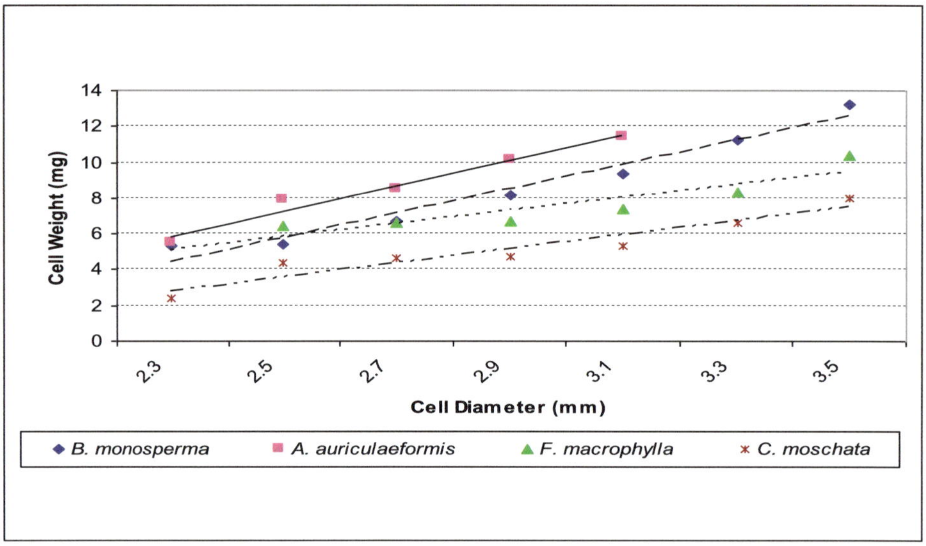

Figure 2.
Resin productivity and cell size relationship in Kerria lacca *(*rangeeni *strain) on different host plants during rainy season crop.*

Host-plant	Initial host preference (%)	% Survival at crop maturity	Diameter of the cell (mm)	Weight diameter ratio	Host suitability index
A. Kusmi strain					
Acacia auriculiformis	26.67	13	3.50	5.404	65.53
Cucurbita moschata	10.00	8	3.02	2.023	4.89
Flemingia macrophylla	80.00	19	3.16	3.617	143.34
Schleichera oleosa	86.67	22	3.54	6.452	435.50
B. Rangeeni strain					
A. auriculiformis	20.00	15	3.10	2.932	27.27
Butea monosperma	93.33	18	3.22	2.720	147.16
C. moschata	13.33	9	3.19	1.881	7.20
Flemingia macrophylla	86.67	18	3.17	2.363	116.85

Table 1.
Host suitability index of different lac host plants for rearing Indian lac insect, Kerria lacca *(Kerr).*

$$\text{Host Suitability} = \text{Index}$$

$$\%\text{Host Preference} \times \%\text{Length of settlement} \times \%\text{lac insect survival}$$
$$\times \text{Mean resin production(mg)} \times 100$$

where % host preference – (% of host-plants on which lac insect survived till crop maturity), % length of settlement at crop maturity – (% length of encrustation of available shoot length), % survival of lac insect (per square cm) at crop maturity –

(% surviving lac insect number of the initial density of settlement) and resin produced (mg) by lac insect – (average weight of resin produced by fifty randomly collected individual female lac insects).

2.3 Local environment

Abiotic and biotic components also affect the suitability of the host plant to lac insect. *Ziziphus xylopyra* (*Ghont*) is a good host in Madhya Pradesh but not in Ranchi (India). Similarly, *C. cajan* and *Grewia* spp. are used as lac hosts in North Eastern states but attempts to cultivate lac *Grewia* at Ranchi proved futile. Analysis of weather data of 1984–2012 of Ranchi (India) revealed that the winter months (December and January) have become colder and pre and post-winter months (November and February) the warmer [22]. These changes in climatic parameters have implications in lac cultivation as it is a critical period of lac insect development (pre-sexual maturity) during the summer season crop. Monsoon and winter rainfall spells and magnitude were also found to affect lac crop performance. Effect of abiotic factors (temperature, rainfall and relative humidity) was correlated with lac production of *rangeeni* crop during 2006–2007 to 2012–2013. It was observed that maximum temperature had a significantly negative (−0.911* and − 0.837*) and RH a positive significant (0.850* and 0.800*) correlation with lac production during the critical crop period (March and April) of development in the summer season (*baisakhi*) crop whereas, during rainy season (*katki*) crop, minimum temperature had a significant negative (−0.765*) correlation with lac production. The vulnerability level of lac insect is high during and prior to sexual maturity stage in the summer crop, thus post-winter season is the critical period for lac insect survival and any undesired variability in weather parameters in this stage can impact adversely on lac productivity.

3. Factors affecting lac quality

Lac insects have been designated as crimson, yellow, cream or white depending upon the quality and intensity of the water-soluble pigment (laccaic acids) present in their body. Resin pigment (erythrolaccin) is not soluble in water and is not desirable in some of its application areas. Quality of the secreted resin due to the presence of dye differed significantly among various stocks of lac insects. *Kusmi* strain of *K. lacca* produced the lightest colored resin whereas that produced by *K. chinensis* and *K. sharda* insects was the darkest. Saha [23] has also reported significant differences in physicochemical properties of lac resin secreted by three different lac insect species. Lac insect species and their host plants also affect the quantity of lac dye (erythrolaccin) present in the particular lac insect stock.

Biochemical parameters of the lac host plants get altered after lac insect inoculation. Fengshu *et al.* [24] have analyzed phenolics, sugars, amino acids and some inorganic elements of the host plants of different ages and strains with and without lac inoculation. The biochemical parameters showed variations after lac inoculation and correlated with the seedlac quality and quantity in different seasons. Liu *et al.* [25] have analyzed tannins and phenolics in relation to the quality and size of the winter generation *K. lacca* on different lac hosts such as *Dalbergia szemaoensis, D. hupeana, D. obtusifolia, Pueraria thunbergiana, Zenia insignis, Cassia siamea* and *C. cajan*. Yang *et al.*, [26] have reported that the host tree root secretion and the quality of the lac produced are the indexes to the inter-adaptability between the lac insect and its host tree such as *Acacia suma, Hibiscus syriacus, Moghania macrophylla and S. oleosa*. Various studies indicated a decrease in the amount yet increase in the variety of amino acids in the host trees on lac inoculation. Lac composition also differed significantly from the lac insects grown on different host trees.

4. Associated fauna

A number of major and minor pests/diseases at times almost destroy the lac crop, thereby, not only reducing the yield drastically but also affecting the quality of the lac. Sessile nature of lac insects makes them more prone to predators and parasitoids.

4.1 Predators

The losses in lac cultivation due to various insect predators are known to be far greater than what is usually met in other agricultural crops. About 22 predators have been reported to be closely associated with lac insects, of which three are major predators *viz*. *Eublemma amabilis*, Moore (Noctuidae); *Pseudohypatopa pulverea* Meyr. (Blastobasidae) and *Chrysopa* spp. (Chrysopidae). *E. amabilis* and *P. pulverea* alone are responsible for 30–40% of damage to the standing crop [27, 28] of which *E. amabilis* alone causes 20–25% damage [29]. These lepidopterous predators cut a hole in the lac and feed on the insect from inside by making a tunnel. *Chrysopa*, though a sporadic pest, sometimes causes havoc, particularly in *kusmi* strain.

Chemical communication between the lac insect-associated products and the lac predators have been evaluated under the laboratory condition and semio-chemicals identified for different stages of lac insect and its associated products [30]. Six major compounds *viz*. decane, dodecane, tetradecane, heptadecane, eicosane and octacosyl acetate constituted 78%, 79% and 85% in lac insect whole body, lac resin and lac wax extracts, respectively. Electroantennogram (EAG) revealed higher responses of adult male and female of lac insect predator, *Eublemma amabilis* to lac insect whole body extract than resin, wax, crawler and lac insect female extracts. Whereas, in *Pseudohypatopa pulverea*, EAG response was significantly higher in females towards lac insect whole body than resin, wax, adult female and crawler extracts than males. Both *E. amabilis* and *P. pulverea* exhibited high level of sensitivity to lac insect whole body extracts with different concentrations ranging from 1000 to 10,000 ppm than the identified semiochemicals *viz*., Decane, Hexadecane, Nonadecane, Eicosane and hexane.

4.2 Parasites

Thirty different parasites of lac insect have been reported by Varshney [31]. They lay eggs into the lac cell through the anal tubercle in/on the body of lac insect. The grub that hatches feeds only on lac insect.

4.2.1 Inimical parasites

Of all the parasites associated with lac insect, eight parasites namely, *Coccophagus tschirchii*, *Erencyrtus dewitzi*, *Eupelmus tachardiae*, *Parechthrodryinus clavicornis*, *Tachardiaephagus tachardiae*, *Marietta javensis*, *T. somervillei* and *A. purpureus* are of regular occurrence in the lac ecosystem. Among these, *Tachardiaephagus tachardiae* and *Aprostocetus purpureus* are the most abundant lac associated parasites. Extent of parasitisation varied between 15.5% in summer season (*baisakhi*) crop and 18.6% in rainy season (*katki*) crop of *rangeeni* strain. While for the *kusmi* strain it was 19.04% in winter season (*aghani*) crop and 22.8% in summer season (*jethwi*) crop [10]. Fecundity of lac insect is adversely affected by parasitisation. Parasitized cells adversely affect the resin production and brood value (fecundity) of the crop. Either, there is no emergence or very low emergence of young ones from the parasitized cells, which ultimately affects the inoculation of the next crop.

Parasitic losses: Percentage of parasitism recorded is much higher than the earlier reports of average 4.8 – 9.9 per cent parasitism based on seven years' data [29]. Reinterpretation of the same data by Srivastava and Chauhan [32], however, revealed that average per cent parasitism for the crop on the basis of females alone worked out to be 20 to 37. In certain years and in some localities, it was as high as 50%. Jaiswal and Saha [33] have found a positive and significant correlation between density of lac insects and number of parasitoids. So it is highly likely that actual per cent parasitism would be higher than recorded since lac insects generally form a continuous encrustation and the present study was confined only to isolated female lac insects.

Rangeeni strain is more vulnerable to pest attack than *kusmi* and damage is more in rainy season crop. Parasites lay eggs into the lac cell through the anal tubercle in or on the body of the lac insect. The grub that hatches feeds only on lac insect and not on lac. As a result of parasitisation, fecundity and resin-producing capability *Kerria lacca* is adversely affected. Quality of the resin produced declined by 17.92% and 17.44% while fecundity decreased by 32.55% and 34.71% for *kusmi* and *rangeeni* strains, respectively [9, 34]. Damage caused by parasites varies depending upon the virulence of outbreak and stage of development of the lac insect at which damage is inflicted. Instances of super-parasitism (as many as 19 larvae have been reported from a single mature female lac insect) and multi-parasitism are not uncommon, which further aggravate the problem.

It was found that the decrease in fecundity of lac insect due to parasitisation ranged between 10–100%. Thus, proportionately more broodlac would be required as compared to healthy broodlac for inoculating the same number of trees. If parasitized at an early stage, the lac insect is practically eaten up by the developing parasitoid rendering the lac useless for broodlac purposes. Moreover, the broodlac harboring parasites if used for raising next crop would serve as a source of infection to new lac culture.

Lac insect is gregarious in nature and resin secreted by these coalesces to form a continuous encrustation. The thickness of lac encrustation is one of the criteria for assessing quality of broodlac. Size of parasitized as well as healthy cells did not differ much but comparison of weight of the resin secreted by the two revealed that the amount of the resin produced by the parasitized cells was significantly lower. Hence, visual assessment of broodlac quality on the basis of encrustation thickness alone may prove to be deceptive unless weight is also taken into account.

4.2.2 Beneficial parasites

Several types of insects are hyper-parasitic to the lac insect. Though their natural population constitutes only about 4–10% of the total fauna associated with lac insect, they act as a bio-control agent in controlling the damage done by inimical insects. Many ants and other insect species feed on the honey dew excreted by lac insect, and these prevent losses by fungus infection.

4.2.3 Effect of fungi on lac production

In addition to the damage caused by insect pests, lac crop yield suffers significant losses due to other biotic agents, particularly fungi. Few earlier reports suggested that the lac insects had mutualistic relationship with fungi. However, an association of fungi with lac insects is not beneficial always. Lac insects being phloem feeders excrete excess sugar in the form of honey dew, which invites sooty mold to grow over the lac encrustation. Besides, this rainy season crop is also prone to fungal infection when grown on *ber* and *kusum* due to their shady nature. Avoidable losses due to fungi alone were observed to be 40.9% to 59.85 in the *kusmi* strain of lac insect [35]. Similarly, Mishra *et al.* [36] have reported significant reduction (75.05%–88.41%) in mortality of second instar lac nymphs with application of different fungicides on the *kusmi* strain of lac insect.

4.2.3.1 Fungi associated with lac insects

The earliest record of honeydew that drips from colonies of lac insects on the twigs of host trees inviting black mold species of *Capnodium* and *Fumago* is that of Lindsay and Harlow [37]; the presence of pathogenic fungi, *pythium* sp. in female tests causes a heavy mortality in the larvae which fail to enclose satisafactorily and lie dead in clusters within the female resinous cell [38]; sooty mold fungi *Conidiocarpus* (Syn. *Podoxyphium conidioxyphium*) and *Polychaeton* spp. are obligate anaerobes capable of producing endospores and causing 30–40% damage to lac insect in Vietnam [39]; 11 species of saprophytic pathogenic fungi causing dark mildew on lac insect have been reported from China [40] and Three species of fungi belonging to family *Eurotiaceae/ Aspergillaceae* causing severe damage to lac culture have been reported from India [41]; *Aspergillus awamori* Nakazawa forms black sheet-like covering on lac encrustation, *Aspergillus terricola* Marchal traverse the whole length of the anal tubercle and blocking it leading to disruption of mating and larval emergence and *Penicillium citrinum* Thom (Syn. *Penicillium aurifluum* biourage) blocks the breathing pores of lac insects. Fungal infection in lac cultures causes losses in lac yield by (i) inhibiting respiration, (ii) hindering mating process, (iii) blocking larval emergence and (iv) affecting lac host efficiency.

5. Interaction of lac insects with microbes

Insects and microbes' interaction ranges from obligate mutualism to facultative parasitism. Insects harbor symbiotic bacteria on the integument, in the digestive tract and in some unique structures within their body [42, 43]. Interaction between insects and microbes is one of the important factors, which makes insects the most successful group of organisms on the earth. Insect microbiota plays significant role in growth, development, reproduction and adaptation of the insects to the environment. Endosymbionts assist insects in their survival by nutritional supplementation [44], aiding in digestion of recalcitrant food materials [45], protecting them from predators and parasitoids [46], detoxifying phytotoxins and pesticides [47], imparting resistance against insecticides [48], immune system stimulation [49], inter and intraspecific communication [50] etc.

Symbiosis with bacterial community is obligatory in insects whose diet is imbalanced such as vertebrate blood (by mosquitoes), phloem sap (by sap-sucking insects) and wood (by termites). Lac insect life stages are morphologically and physiologically highly diverse. Besides, their resin production potential also varies with different stages; crawlers and adult males do not secrete resin whereas, settlers and adult females after fertilization secrete resin albeit in different quantities. Lac insects are almost sedentary throughout their life and depend on nutritionally imbalanced sugar-rich phloem sap for their survival. It is postulated that lac insects must harbor myriad of endosymbionts for nutrition supplementation, host plant adaptability, defense etc. Earlier works have found that the presence of microbial flora in lac insects was beneficial during rainy season crops for higher lac yield. Some bacteria such as *Micrococcus* spp., *Clostridium* sp. and *Bacillus subtilis* have been reported from lac insects [51]. *Micrococcus varians* and *Micrococcus conglomerates* are associated with various stages of lac insect and are considered beneficial for good yield of lac production.

Till now, bacterial flora associated with lac insects has been identified either based on culturing method [52, 53] or PCR method for specific endosymbionts like *Wolbachia* [54]. Gender-specific bacterial flora has been identified from lac insects [53], and host plant-induced variation was observed in the bacterial composition of lac insects based on culturing method [52].

5.1 Association with *Wolbachia*

Wolbachia are members of the order Rickettsiales, a diverse group of intracellular bacteria that include species having parasitic, mutualistic and commensal relationships with their hosts. *Wolbachia* species are well known for their vast abundance, effect on hosts in terms of reproductive manipulation and mutualism and have potential applications in pest and disease vector control [55]. *Wolbachia pipientis* is the type species of *Wolbachia* genus. Based on the 16S ribosomal sequence and other sequence information, *Wolbachia* spp. have been divided into seventeen different supergroups (A-Q). Two supergroups (C and D) are commonly found in filarial nematodes, whereas other groups are found in arthropods, in which A and B are the most common. *Wolbachia* that participate in symbiotic relationships with arthropods have a range of phenotypic effects on their hosts and generally behave as reproductive parasites. *Wolbachia* manipulate host reproduction through cytoplasmic incompatibility, parthenogenesis, feminization and male-killing [56–58]. Mostly *Wolbachia* undergoes vertical transmission from mother to offspring. However, a horizontal transfer is also reported in nature [58].

In lac insect populations, there is a wide variation of male–female sex ratio, which ultimately affects the lac production as only females can produce lac. Vashishtha and co-workers [54] have found that lac insects are associated with *Wolbachia* based on 16S rDNA and wsp (*Wolbachia* cell surface protein) PCRs. Lac insect-associated *Wolbachia* was termed as wKerlac. Phylogenetic tree revealed it to be a subgroup "ori" of supergroup B, which is predominantly present in arthropods. *Wolbachia* of *K. lacca* was grouped with *Wolbachia* of *Tagosodes orizicolus* and *Ephestia cautella*. *Wolbachia* on both these hosts are responsible for cytoplasmic incompatibility. Further investigations are required on whether the identified *Wolbachia* would have any role in feminization. It is one of the most important factors in attributing lac yield because commercial lac is obtained solely from female lac insects.

5.2 Detection of *Wolbachia* phage (WO) in lac insects

Wolbachia species also harbor a bacteriophage called bacteriophage WO or phage WO [59]. Comparative sequence analyses of bacteriophage WO revealed the possibility of large-scale horizontal gene transfer between *Wolbachia* coinfections in the same host [60]. Molecular mechanism used by *Wolbachia* to manipulate its host in terms of cytoplasmic incompatibility, feminization, parthenogenesis, male killing *etc.* remain elusive and has been speculated due to genes on extrachromosomal factors such as plasmids or bacteriophages [61, 62]. Out of seventeen identified super-groups of *Wolbachia*, named A–Q, WO phage has been reported to infect *Wolbachia* belonging to super-group A, B, F and G [63, 64].

Screening and distribution of *Wolbachia* and WO phage sequences were studied by amplifying and sequencing the partial *ftsZ*, a cell cycle gene involved in cell division and a putative minor capsid protein, *orf* 7, respectively [65]. Two different lines *kusmi* and *rangeeni* each were found to be singly infected by *Wolbachia* belonging to Super-group B. It was the first report on molecular detection of WO-phage infecting *kusmi* and *rangeeni* infrasubspecific forms of *K. lacca*. Further phylogenetic analysis revealed distinct differentiation of WO between *kusmi* and *rangeeni* infrasubspecific forms. In the phylogenetic tree made based on *orf7* sequences, *rangeeni* and *kusmi* forms clustered with group III and group I, respectively. Although there is a differentiation of *kusmi* and *rangeeni* forms based on *orf7* of WO sequences, the tripartite association of lac insect-*Wolbachia*-WO needs further investigation to implicate their role in such differentiation.

5.3 Association with yeast-like symbionts

Insects not only possess bacterial symbionts but also yeast-like fungal symbionts (YLS). Although microbiologists observed such yeast-like endosymbionts in 1960s, the identity was not known due to their fastidious nature and the lack of molecular tools at that time. Later, it was found that the phloem sap feeders harbor obligate intracellular yeast-like symbionts, YLS (subphylum Ascomycota, class Pyrenomycetes, family Clavicipitaceae) [66, 67] especially in the mycetocytes formed by fat body cells of abdomen. YLS have also been reported in Hemiptera (aphids, planthoppers, and scale insects) and Coleoptera (beetles) [68, 69]. They grow by budding and are vertically transmitted to the next generation by transovarial infection [70, 71]. YLS appears to play roles in nitrogen metabolism of the host through recycling of uric acid [72, 73], in insect metabolism by synthesizing sterols, the precursor molecule for many hormones (*e.g.*, 20-OH ecdysone-a molting hormone), as insects, in general, are unable to synthesize them [74] and in detoxifying the toxic substances to the host [75]. Occurrence of YLS is highly essential for the survival and reproduction of the host insects because they play vital role in development, reproduction and embryonic development [76]. Transmission electron microscopy and PCR-based studies revealed the presence of YLS in lac insects. However, acquisition of YLS in lac insects seems to be different from that of aphids and plant hoppers and a horizontal transfer was also suggested for them [54].

5.4 Sex-specific endosymbionts

Male and female lac insects' specific bacterial species were identified by Shamim and co-workers [53] based on 16S rDNA PCR and biochemical characterization. Eight different bacterial species were isolated and categorized as endosymbiont, gut bacterium or subsurface bacterium. Three of them were exclusive to males, three to females and two were common to both the sexes (**Table 2**). *Bacillus megaterium*, *A. subterraneus* and *Pantoea ananatis* were found to be the most abundant bacterial species. Among 13 bacterial isolates found in males, two were present as an internal gut bacterium which may get excreted out with honey dew. Single isolate of *Paenibacillus barengoltzii* was found at subsurface. *P. ananatis* was exclusively and majorly found in males as endosymbiont bacteria. *P. fulva* was also found in males as endosymbiont bacteria. Twelve bacterial species were isolated from the females and 50% populated with *Bacillus* sp. *B. megaterium, Curtobacterium citreum* and *A. subterraneus* were majorly reported in non-crushed samples, therefore, it was thought to be thriving at subsurface but *B. cereus* and *Solibacillus silvestris* were considered as endosymbiont. *A. subterraneus,* found in both the sexes, as endosymbiont in males and at subsurface in females.

Out of these bacterial species, *B. cereus, B. megaterium* and *P. ananatis*, are widespread in occurrence and have also been reported in other insect's body. Some strains of *P. ananatis,* also referred as "ice nucleation-active" bacteria, are used in pest control because when present in the insect gut, they lower the cold resistance [77]. Owing to this property and bacterial abundance in the insect, absence of lac insects in colder regions could be attributed to the association of *P. ananatis*. Description about the bacterial species identified from both the sexes of lac insect [53] is given in **Table 2**.

5.5 Host plant-induced variation of endosymbionts

Since phloem sap constituents vary for host plants, variation in the endosymbionts of lac insects growing on different host plants is anticipated. Culture-based method was followed to isolate bacteria from lac insects grown on different host plants and 16S rDNA PCR based molecular method was followed [52] to identify them.

From 29 different bacterial isolates, 10 different bacteria were identified. *Bacillus kochii, Bacillus oceanisediminis, Bacillus amyloliquefaciens, Bacillus nakamurai* and *Enterobacter cloacae* were observed on *kusmi* lac insects collected from *Kusum* trees. *Klebsiella quasipneumoniae subsp. similipneumoniae, Citrobacter amalonaticus,* P*rovidencia vermicola* and *B. nakamurai* were found in bacteria isolated from lac insects collected from *ber* trees. *Enterobacter ludwigii, Enterobacter cancerogenus* and *B. nakamurai* were found in lac insects collected from *semialata*. In most of the cases, different species of *Bacillus* and *Enterobacter* were found. *Bacillus* is a very common genus found in different types of insects, which include *B. subtilis, Bacillus thuringiensis, Bacillus cereus, Bacillus sphaericus, Bacillus popillae, Bacillus circulans, B. megaterium, Bacillus lentimorbus a*nd *Bacillus polymyxa* [78]. Bacteria such as *Enterobacter* spp. *K. quasipneumoniae, C. amalonaticus, P. vermicola* belong to Enterobacteriaceae family. All these Enterobacters might have come from lac insect gut and belong to proteobacteria, primarily within the γ–subdivision.

B. nakamurai the most frequent bacteria identified from lac insects grown on three different hosts might be involved in some vital functions in lac insects. *B. nakamurai* was originally isolated from soil and known to produce black pigment [79].

Associated with	Bacterial spp.
Exclusively male lac insects	*Paenibacillus barengoltzii, Pseudomonas fulva* and *Pantoea ananatis*
Exclusively female lac insects	*Bacillus cereus, Solibacillus silvestris* and *Curtobacterium citreum*
Both male and female lac insects	*Bacillus megaterium* and *Arthrobacter subterraneus*

Table 2.
Bacterial species identified from lac insects.

B. amyloliquefaciens is known to control plant pathogens due to its antifungal activity [80]. Hence, antifungal activity may be anticipated for the *B. amyloliquefaciens* strain present in lac insects.

6. Conclusion

6.1 Impact on quantity and quality

Lac insects feed on host plant phloem sap by passive sucking mechanism through capillary action and turgor pressure of the sap [81]. Passive exudation of phloem sap through the intruded lac insect stylet as a function of phloem turgor pressure has a significant role in the supply of phloem sap to lac insects. Lac insects are sedentary soon after settlement on host plants. As a result, there is no escape mechanism for lac insects from the host plant's defensive chemicals. Hence, lac insects should have ability to detoxify the host plant's defensive chemicals and avoid them which is important for good lac insect – host plant interaction. Due to differences in the ability of lac insects to detoxify the defensive chemicals and the nature of defensive chemicals produced by the plants, there is difference in yield of lac on different lac host plant species and different genotypes of the same host plant. Lesser quantities of toxicant ingested when feeding on phloem sap of good host plants rich in nutrients would be detoxified quickly, whereas the larger quantities of toxicant ingested when feeding on poor quality host plant may not be detoxified by lac insects. Haque (1984) observed variations in the quality and quantity of amino acids present in honey dew (anal fluid) of *K. lacca* grown on different host plants such as *Moghania* (=*Flemingia*) *macrophylla*, *Ficus glomerata*, *Ficus indica* and *Ficus religiosa* [82], indicating the variations in nutritional composition of these host plants.

Due to sub-optimal nitrogen/carbon ratio in the phloem sap, phloem feeders need to ingest excess phloem sap along with excess organic carbon to obtain sufficient nitrogen for their growth. Non-optimal ratios of essentially required nutrients in the phloem sap warrant the insects to ingest supra optimal quantities of less required nutrients and also the defensive chemicals, which are toxic to them. These factors affect the resin-producing efficiency of lac insects. Resin production is found to be high on tree hosts in comparison with bushy hosts such as *F. macrophylla* and pumpkin fruit. Auclair [83] has reported that the phloem sap exudation rate through excised stylets in aphids is about an order of magnitude higher in woody plants compared to that in herbaceous plants. Since lac insect feeding habit is also similar to aphids and passive in nature, the higher turgor pressure of woody plants leads to greater food ingestion and higher resin production on tree hosts and vice versa in herbaceous hosts. Intraspecific variation in the host plant defense caused the deviation in the susceptibility of a host plant species in *Nuculaspis californica* [84]. Lac insects have the ability to manifest their biological parameters very specifically not only to lac host species but also to varieties, phenotypes of host plants [18, 85] locality and season of cultivation. Periodic host-plant resistance is considered to be a physiological response to meteorological and edaphic conditions, a response usually rendering the plant temporarily unsuitable to coccoid development.

It can be concluded that lac insects, as well as lac-host plants and associated flora and fauna, play significant roles in the quantity and quality of the lac production. The naturally existing high degree variability in lac insects and host plants can be successfully exploited for selection and evolution of high-yielding varieties and lac insect-host combinations. Quantity and quality of the lac resin can be significantly enhanced with better management of lac insects and their host plants.

6.2 Interaction with microbes

Crawlers *i.e.*, early developmental stage of lac insect consists more of unknown and other bacterial types followed by *Wolbachia* and *Mucilaginibacter*. *Wolbachia* and *Pantoea* are the two important genera found in the adult female lac insects besides unknown bacteria. *P. ananatis* is already reported to be present exclusively in male lac insects based on the culture method [53]; *Pantoea cypripedii* and *Pantoea dispersa* are present in honeydew secreted by lac insects [86]. In insects, *Pantoea* is primary endosymbiont and its association with insects is mostly mutualistic and sometimes commensalistic. In the mutualistic association, insects provide habitat and nutrition to *Pantoea*, whereas *Pantoea* may help insects by hydrolysis of proteins, antagonism of pathogens, breakdown of toxic substances, nitrogen fixation, nutrition and digestion [87]. Since *Pantoea carbekii* genome encodes complete or near-complete canonical pathways for the production of several vitamins and cofactors such as folate, riboflavin, pyridoxal-5′-phosphate, glutathione, iron–sulfur clusters and lipoate, it is assumed that *Pantoea* supplements nutrition by providing essential vitamins and minerals to its host stink bug [88]. Taking these things into account, it is plausible to assume that the *Pantoea* spp. present in lac insects may be involved in nutrition supplementation because plant phloem sap on which lac insects feed is not a nutritionally balanced diet.

Wolbachia is an obligate endosymbiont of arthropods and nematodes and present in most of the insects wherein they play a major role in the reproductive manipulation of the host. They alter the reproduction of the host insects by the way of male killing, parthenogenesis, cytoplasmic incompatibility and feminization. Generally, the mode of transfer of *Wolbachia* in arthropods and nematodes from one generation to another is transovarial [89]. Due to such reproductive manipulation of the host by *Wolbachia*, the frequency of *Wolbachia* infected females increases in population sometimes at the expense of host fitness [90]. Vashishtha and co-workers [54] have reported the presence of *Wolbachia* in lac insects by 16S rDNA and *wsp* PCRs. *Wolbachia* infection is known to be biased based on sex or may increase as the development progress. Frequency of occurrence of *Wolbachia* was found to be more in female insects compared to crawlers in the current study. Similar results of higher *Wolbachia* incidence in the adult stage compared to an immature stage were obtained in several other insects. Besides manipulating host reproduction, *Wolbachia* may affect host fitness positively by nutrient supplementation. It has been demonstrated in bed bugs, *Cimex lectularius* that riboflavin provision ability of *Wolbachia* can positively impact the host's growth, survival and reproduction [91]. In the scale insect, *Dactylopius coccus*, two species of *Wolbachia* were found to have metabolic capabilities for riboflavin and heme biosynthesis [92]. Besides, reproductive manipulation and nutrient supplementation, an additional function of protecting *Drosophila* from virus induced mortality was attributed by *Wolbachia* infection [93].

As far as lac insects are concerned, females are the productive gender as the commercial lac is obtained only from females but not from male insects. Since, *Wolbachia* can eliminate males, turn them into females, sterilize uninfected females or behave as a mutualistic symbiont [94], their role in lac production needs to be explored thoroughly. Whether the role of *Wolbachia* is restricted up to reproductive manipulation or it is extended to nutrient supplementation and virus protection in lac insects needs thorough investigation in future.

Lac insect endosymbionts are very diverse and supposed to carry out various vital functions in the insects. The available literature describe mainly cultivable bacteria and to certain extent uncultivable microbes. Much more uncultivable bacteria and also stage-dependent and strain-dependent endosymbionts may be anticipated to be present in lac insects. Culture-independent methods such as metagenomics would reveal more number of endosymbionts in lac insects. Different functions such as nutrition supplementation, sex differentiation, strain differentiation and protection from pests and predators are anticipated for the lac insect endosymbionts.

Author details

Kewal Krishan Sharma* and Thamilarasi Kandasamy
ICAR-Indian Institute of Natural Resins and Gums, Namkum, Ranchi, Jharkhand, India

*Address all correspondence to: kewalkks@gmail.com

References

[1] Siddiqui SA. Lac-the versatile natural resin. Natural Product Radiance. 2004;**3**: 332-337

[2] Ben-Dov Y, Lit IL. Stabilizing Kerriidae as the family-group name of the lac insects (hem. Coccoidea). Bulletin de la Sociéte entomologique de France. 1998;**103**(5):455-456

[3] Varshney RK. Tachardiidae versus Kerriidae as family name of the lac insects (Hemiptera: Coccoidea). Bionotes. 1999;**1**(3):61-62

[4] Varshney RK. In: Sharma KK, editor. Lac Insects of the World – An Updated Catalogue and Bibliography. Ranchi, India: ICAR-IINRG; 2020. p. viii + 84

[5] Talukder B, Das BK. A new invasive species of *Kerria* Targioni-Tozzetti (Hemiptera: Coccomorpha: Kerriidae) on rain tree, *Albizia saman* (Fabaceae) from India. Oriental Insects. 2021;**55**:176-190. DOI: 10.1080/00305316.2020.1759467

[6] Rajgopal NN, Mohanasundaram A, Sharma KK. A new species of lac insect in the genus Kerria Targioni Tozzetti (Hemiptera: Coccomorpha: Tachardiidae) on *Samanea saman* (Fabaceae) from India. Zootaxa. 2021; **4938**(1):60-68. DOI: 10.11646/zootaxa.4938.1.2

[7] García, Morales M, Denno BD, Miller DR, Miller GL, Ben-Dov Y, Hardy NB. ScaleNet: A literature-based model of scale insect biology and systematics. Database. 2016

[8] Jaiswal AK, Sharma KK. Lac insect its development and life cycle. In: Kumar KK, Ramani R, Sharma KK, editors. Recent Advances in Lac Culture. Ranchi, India: Indian Institute of Natural Resins and Gums; 2011. pp. 22-26

[9] Sharma KK, Kavita K, Suman L. Host plant mediated variations in resin producing efficiency of Indian lac insect, *Kerria lacca* (Kerr) (Homoptera: Coccoidea: Tachardiidae). Entomon. 2007;**32**(3):203-207

[10] Kavita K. Host-Plant Mediated Variation in Productivity Linked Attributes of Indian Lac Insect, *Kerria lacca* [Thesis]. Hazaribagh: Vinoba Bhave University; 2011

[11] Glover PM. Lac Cultivation in India. Namkum, Ranchi: Indian Lac Research Institute; 1937. p. 147

[12] Chauhan NS. Sex in relation to sequence of larval emergence. Indian Journal of Entamology. 1970;**32**(3): 273-274

[13] Krishnaswami S, Purkayastha BK, Majumdar N. Studies on the determination of optimum amount of brood lac requirement for crop inoculations. Part I for Kusum (*Schleichera oleosa*) in *jethwi* and *aghani* seasons. Indian Forester, Dehradun. 1963;**89**(8):540-549

[14] Purkayastha BK, Krishnan YS. Determination of optimum amount of broodlac requirement for crop inoculation and proper time of harvesting for maximizing yield on *palas* in the *Baisakhi* season. Indian Agriculturist, Calcutta. 1964;**8**(1):55-60

[15] Chauhan NS. Studies on Sex Determination in Lac Insects. Annual Report. Ranchi: Indian Lac research Institute; 1988. pp. 11-14

[16] Srivastava DC, Kumar P. Sex ratio of the Indian lac insect on pigeon-pea. Legume Research. 1985;**8**(2):101-102

[17] Choudhary MK, Gokulpure RS, Bhattacharya A. Efficacy of Different Host-Plants on the Survival of Lac Insects. Annual Report 1974–76. Ranchi (India): Indian Lac Research Institute; 1982. pp. 6-9

[18] Srinivasan MM. Lac, its cultivation and manufacture in India, chapter IV - host plants of the lac insect. Indian Forester. 1956;**82**(4):180-193

[19] Sandeep K, Amit V, Shweta S, Sharma KK, Suman L. Essential amino acid profiling of the four lac hosts belonging to genus *Flemingia*: Its implications on lac productivity. Physiology and Molecular Biology of Plants. 2020;**26**:1867-1874. DOI: 10.1007/s12298-020-00860-9

[20] Thakur AK. Comparative Study of Lac Hosts with Special Reference to *Acacia Catechu* and Cassia Florida. Vol. 9. Namkum, Ranchi: Bull. Indian Lac. Res. Inst; 1932. p. 1

[21] Sharma KK, Ramani R. Suitability of pumpkin (*Cucurbita moschata* Duchense ex Poir) fruits for laboratory rearing of two strains of Indian lac insect, *Kerria lacca* (Kerr.) (Coccoidea: Tachardiidae). Journal of the Entomological Research. 1997;**21**(2):169-174

[22] Mohanasundaram A, Monobrullah M, Sharma KK, Anees K, Singh RK, Meena SC, et al. Climate change: Effect of weather parameters on production of summer season crop of *rangeeni* strain of Indian lac insect, *Kerria lacca* (Kerr) at Ranchi, Jharkhand. Journal of Agrometeorology. 2014;**16** (Special Issue-I):108-113

[23] Saha SK. Physiochemical study of lac from three insect species. Tropical Science. 1995;**35**:13-16

[24] Fengshu L, Yi P, Li S. The relation of the property of lac of lac insect to the chemical composition of the host plant. Scientia-Silvae-sinicae. 1988;**24**:106-112

[25] Liu FS, Yi P, Li SJ. Relation of lac production by lac insects to the chemical composition of host plants. Scientia Silvae Sinicae. 1988;**24**(1):106-112

[26] Yang CY, Zhang JY, Cai J, Lu WH, Deng J, Chen YH. A study on the adaptability response between the lac insect and its host trees. Forest Research. 1998;**11**:647-651

[27] Malhotra CP, Katiyar RN. Chemical control of the lac predator *Eublemma amabilis* Moore II. Relative toxicity of the various insecticides against the predatory caterpillars. Indian Entomology. 1979;**41**:187-190

[28] Bhattacharya A, Jaiswal AK, Kumar S, Kumar M. Evaluation of cartap hydrochloride for management of *Eublemma amabilis* (Moore) - a serious lepidopteron predator of lac insect. Journal - Applied Zoologist Research Association. 2005;**16**:93-94

[29] Narayanan ES. Pests of lac in India. In: Mukhopadhyay B, Muthana MS, editors. A Monograph on Lac. Namkum, Ranchi: Indian Lac Research Institute; 1962. pp. 90-113

[30] Arumugam M, Kaprakkaden A, Nebapure SM, Sharma KK, Naaz N, Mishra R, et al. Electroantennography and behavioral studies of *Eublemma amabilis* [Moore] and *Pseudohypatopa pulverea* [Meyr] in relation to volatiles of lac insect (*Kerria lacca* Kerr.) and its associated products. International Journal of Tropical Insect Science. 2022; **42**:2313-2324. DOI: 10.1007/s42690-022-00754-1

[31] Varshney RK. A check-list of insect parasites associated with lac. Oriental Insects. 1976;**10**(1):55-78

[32] Srivastava DC, Chauhan NS. A critical appraisal of parasitic losses in lac. Indian Shellac. 1984;**24**(1 and 2):1-2

[33] Jaiswal AK, Saha SK. Estimation of the population of parasitoids associated with lac insect, *Kerria lacca* (Kerr) on the basis of biometrical characters. Journal of the Entomological Research Society. 1995;**19**(1):27-32

[34] Krishan SK, Ramani R. Parasites effected reduction in fecundity and resin yield of two strains of Indian lac insect, *Kerria lacca* (Kerr). Indian. Journal of Entomology. 2001;**63**(4):456-459

[35] Das BB, Naqvi AH, Choudhary SG, Sen AK. Survey of Pathogenic Microorganisms and their Effects on Development of Lac Insects. Annual Progress Report. Namkum, Ranchi: Indian Lac Research Institute; 1986. pp. 24-25

[36] Mishra YD, Choudhary SG, Singh BP, Mishra RC, Naqvi AH, Kumar P. Advance in lac production technology. In: National Seminar on Lac Industry – Challenges and Solution. Ranchi: ILRI; 1997; 14-18 June, 1996 at. pp. 33-42

[37] Lindsay HAF, Harlow CM. Report on Lac and Shellac. Allahabad, India: Indian Forest Records; 1921. pp. 8-162

[38] Misra CS. The Cultivation of Lac in the Plains of India (*Laccifer Lacca* Kerr.). Calcutta, India: Agricultural Research Institute; 1928. Pusa Bulltein No. 185

[39] Malhotra CP. Lac insect complex of the world, In: An Exposition of Status and Development of Lac Technology in India. Indian Lac Research Institute, Ranchi. 1991; Pp. I-3 to II-11.

[40] Shaoji G. An investigation on the pathogens of both lac insects and host plants. Forest Research. 1993;**6**:711-713

[41] Krishan SK, Jaiswal AK, Kumar KK. New record of fungi associated with Indian lac insect, *Kerria lacca* (Kerr). Indian. Journal of Entomology. 2001; **63**(3):369-371

[42] Chen DQ, Montllor CB, Purcell AH. Fitness effects of two facultative endosymbiotic bacteria on the pea aphid, *Acyrthosiphon pisum*, and the blue alfalfa aphid, A. kondoi. Entomologia Experimentalis et Applicata. 2000;**95**: 315-323

[43] Fukatsu T, Nikoh N, Kawal R, Koga R. The secondary endosymbiotic bacterium of the pea aphid *Acyrthosiphon pisum* (Insecta: Homoptera). Applied and Environmental Microbiology. 2000; **66**:2748-2758

[44] Muhammad A, Fang Y, Hou Y, Shi Z. The gut Entomotype of red palm weevil *Rhynchophorus ferrugineus* Olivier (Coleoptera: Dryophthoridae) and their effect on host nutrition metabolism. Frontiers in Microbiology. 2017;**8**:2291. DOI: 10.3389/fmicb.2017.02291

[45] Zhou J, Duan J, Gao M, Wang Y, Wang X, Zhao K. Diversity, roles, and biotechnological applications of symbiotic microorganisms in the gut of termite. Current Microbiology. 2018;**76**:755-761. DOI: 10.1007/s00284-018-1502-4

[46] Oliver KM, Russell JA, Moran NA, Hunter MS. Facultative bacterial symbionts in aphids confer resistance to parasitic wasps. Proceedings of the National Academy of Sciences. 2003; **100**:1803-1807. DOI: 10.1073/pnas.0335320100

[47] Itoh H, Tago K, Hayatsu M, Kikuchi Y. Detoxifying symbiosis: Microbe-mediated detoxification of phytotoxins and pesticides in insects. Natural Product Reports. 2018;**35**: 434-454. DOI: 10.1039/c7np00051k

[48] Kikuchi Y, Hayatsu M, Hosokawa T, Nagayama A, Tago K, Fukatsu T. Symbiont-mediated insecticide resistance. Proceedings of the National Academy of Sciences. 2012;**109**:8618-8622. DOI: 10.1073/pnas.1200231109

[49] Kwong WK, Mancenido AL, Moran NA. Immune system stimulation by the native gut microbiota of honey bees. Royal Society Open Science. 2017; **4**:170003. DOI: 10.1098/rsos.170003

[50] Calcagnile M, Tredici SM, Talà A, Alifano P. Bacterial Semiochemicals and Transkingdom interactions with insects and plants. Insects. 2019;**10**:441. DOI: 10.3390/insects10120441

[51] Sharma KK, Jaiswal AK. Microbial flora associated with lac insects. In: Kumar KK, Ramani R, Sharma KK, editors. Recent advances in lac culture. Ranchi, India: Indian Institute of Natural Resins and Gums; 2011. pp. 112-115

[52] Thamilarasi K, Sajiya E, Monika G, Kanchan K, Lohot Vaibhav D, Mohanasundaram A, et al. Plant host induced variation of endosymbionts associated with *Kusmi* lac insects. Multilogic in Science. 2018;**VIII**(C):99-102

[53] Shamim G, Ranjan SK, Kandasamy T, Sharma KK, Ramani R. Bacterial flora associated with *Kerria lacca* (Kerr). Indian Journal of Entomology. 2017;**79**:41-48. DOI: 10.5958/0974-8172.2017.00010.4

[54] Vashishtha A, Sharma KK, Lakhanpaul S. Co-existence, phylogeny and putative role of Wolbachia and yeast-like symbiont (YLS) in *Kerria lacca* (Kerr). Current Microbiology. 2011;**63**:206-212. DOI: 10.1007/s00284-011-9961-x

[55] Bourtzis K. Wolbachia-based technologies for insect pest population control. Advances in Experimental Medicine and Biology. 2008;**627**:104-113

[56] O'Neill SL, Hoffmann AA, Werren JH. Influential Passengers: Inherited Microorganisms and Arthropod Reproduction. Oxford: Oxford University Press; 1997

[57] Rousset F, de Stordeur E. Properties of *Drosophila simulans* strains experimentally infected by different strains of the bacterium Wolbachia. Heredity. 1994;**72**:325-331

[58] Werren JH. Biology of Wolbachia. Annual Review of Entomology. 1997;**42**: 587-609

[59] Masui S, Kamoda S, Sasaki T, Ishikawa H. Distribution and evolution of bacteriophage WO in Wolbachia, the endosymbiont causing sexual alterations in arthropods. Journal of Molecular Evolution. 2000;**51**:491

[60] Kent BN, Salicos L, Gibbons JG, Rokas A, Newton ILG, Clark ME, et al. Complete bacteriophage transfer in a bacterial endosymbiont (Wolbachia) determined by targeted genome capture. Genome Biology and Evolution. 2011;**3**: 209-218

[61] Bordenstein SR, Wernegreen JJ. Bacteriophage flux in endosymbionts (Wolbachia): Infection frequency, lateral transfer and recombination rates. Molecular Biology and Evolution. 2004;**21**: 1981-1991

[62] Wu M, Sun LV, Vamathevan J, Riegler M, Deboy R, Brownlie JC, et al. Phylogenomics of the reproductive parasite Wolbachia pipientis wMel: A streamlined genome overrun by mobile genetic elements. PLoS Biology. 2004;**2**: 0327-0341

[63] Gavotte L, Vavre F, Henri H, Ravallec R, Stouthamer R, Bouletreau R.

Diversity distribution and specificity of WO phage infection in Wolbachia of four insect species. Insect Molecular Biology. 2004;**13**:147-153

[64] Wang GH, Jia LY, Xiao JH, Huang DW. Discovery of a new Wolbachia supergroup in cave spider species and the lateral transfer of phage WO among distant hosts. Infection, Genetics and Evolution. 2016;**41**:1-7

[65] Kaushik S, Sharma KK, Ramani R, Lakhanpaul S. Detection of Wolbachia phage (WO) in Indian lac insect [*Kerria lacca* (Kerr.)] and its implications. Indian Journal of Microbiology. 2018; **59**(2):237-240. DOI: 10.1007/s12088-018-0763-8

[66] Noda H, Nakashima N, Koizumi M. Phylogenetic position of yeast like symbiotes of rice plant hoppers based on partial 18S rDNA sequences. Insect Biochemistry and Molecular Biology. 1995;**25**:639-646

[67] Suh SO, Noda H, Blackwell M. Insect symbiosis: Derivation of yeast-like endosymbionts within an entomopathogenic filamentous lineage. Molecular Biology and Evolution. 2001; **18**:995

[68] Buchner P. Endosymbiosis of Animals with Plant Microorganisms. New York: Inter Science/Wiley Publishers; 1965

[69] Jurzitza G. The fungi symbiotic with anobiid beetles. In: Batra LR, editor. Insect–Fungus Symbiosis. Montclair: Allanheld, Osmun & Co.; 1979. pp. 65-76

[70] Cheng DJ, Hou RF. Histological observations on transovarial transmission of a yeast-like symbiote in *Nilaparvata lugens* Stål (Homoptera: Delphacidae). Tissue & Cell. 2001;**33**: 273-279

[71] Michalik A, Jankowska W, Szklarzewicz T. Ultrastructure and transovarial transmission of endosymbiotic microorganisms in *Conomelus anceps* and *Metcalfa pruinosa* (Insecta: Hemiptera: Fulgoromorpha). Folia Biologica (Kraków). 2009;**57**:131-137

[72] Sasaki T, Kawamura M, Ishikawa H. Nitrogen recycling in the brown plant hopper, *Nilaparvata lugens*: Involvement of yeast-like endosymbionts in uric acid metabolism. Journal of Insect Physiology. 1996;**42**:125-129

[73] Hongoh Y, Ishikawa H. Uric acid as a nitrogen resource for the brown plant hopper, *Nilaparvata lugens*: Studies with synthetic diets and aposymbiotic insects. Zoological Science. 1997;**14**:581-586

[74] Robbins WE, Kaplanis JN, Svoboda JA, Thompson MJ. Steroid metabolism in insects. Annual Review of Entomology. 1971;**16**:53-72

[75] Dowd PF. In situ production of hydrolytic detoxifying enzymes by symbiotic yeasts in the cigarette beetle (Coleoptera: Anobiidae). Journal of Economic Entomology. 1989;**82**:396-400

[76] Lee YH, Hou FR. Physiological roles of yeast like symbiote in reproduction and embryonic development of the brown plant hopper, *Nilaparvata lugens*, Stål. Journal of Insect Physiology. 1987; **33**:851-860

[77] Coutinho TA, Venter SN. *Pantoea ananatis*: An unconventional plant pathogen. Molecular Plant Pathology. 2009;**10**:325-335

[78] Ince IA, Demir I, Demirbag Z, Nalcacioglu R. A cytoplasmic polyhedrosis virus isolated from the pine processionary caterpillar, *Thaumetopoea pityocampa*. Applied Microbiology and Biotechnology. 2008;**17**:632-637

[79] Dunlap CA, Saunders LP, Schisler DA, Leathers TD, Naeem N, Cohan FM, et al. *Bacillus nakamurai* sp. nov., a black-pigment-producing strain. International Journal of Systematic and Evolutionary Microbiology. 2016;**66**(8):2987-2991

[80] Nam HS, Yang HJ, Oh BJ, Anderson AJ, Kim YC. Biological control potential of *bacillus amyloliquefaciens* kb3 isolated from the feces of Allomyrina dichotoma larvae. The Plant Pathology Journal. 2016;**32**(3):273-280

[81] Sreenivasaya M. Contribution to the scientific study of the lac. Journal of Indian Institute of Science. 1924;**7**(7):120-125

[82] Haque MS. Free amino acids in the anal fluid of female lac insect (*Kerria lacca*) feeding on different host plants. Indian Journal of Entomology. 1984; **46**(3):291-298

[83] Auclair JL. Aphid feeding and nutrition. Annual Review of Entomology. 1963;**8**:439-490

[84] Edmund GF Jr, Alstad DN. Coevolution in insect herbivores and conifers. Science. 1978;**199**:941-945

[85] Mishra YD, Sushil SN, Bhattacharya A, Kumar S, Mallick A, Sharma KK. Intra specific variation in host-plant affecting productivity of Indian lac insect, *Kerria lacca* (Kerr). Journal of Non-Timber Forest Products. 1999;**6**(3/4):114-116

[86] Shamim G, Sharma KK, Ramani R. Isolation and identification of culturable bacteria from honeydew of Indian lac insect, *Kerria lacca* (Kerr) (Hemiptera: Tachardiidae). Meta Gene. 2019;**19**:10-14. DOI: 10.1016/j.mgene.2018.09.010

[87] Walterson AM, Stavrinides J. Pantoea: Insights into a highly versatile and diverse genus within the Enterobacteriaceae. FEMS Microbiology Reviews. 2015;**39**:968-984

[88] Kenyon LJ, Meulia T, Sabree ZL. Habitat visualization and genomic analysis of "*Candidatus Pantoea carbekii*," the primary symbiont of the Brown Marmorated stink bug. Genome Biology and Evolution. 2015;**7**:620-635. DOI: 10.1093/gbe/evv006

[89] Correa CC, Ballard JWO. Wolbachia associations with insects: Winning or losing against a master manipulator. Frontiers in Ecology and Evolution. 2016;**3**:153. DOI: 10.3389/fevo.2015.00153

[90] Kikuchi Y. Endosymbiotic bacteria in insects: Their diversity and Culturability. Microbes and Environments. 2009;**24**:195-204. DOI: 10.1264/jsme2.me09140s

[91] Moriyama M, Nikoh N, Hosokawa T, Fukatsu T. Riboflavin provisioning underlies Wolbachia's fitness contribution to its insect host. MBio. 2015;**6**:e01732-e01715. DOI: 10.1128/mbio.01732-15

[92] Ramírez-Puebla ST, Ormeño-Orrillo E, Vera-Ponce de León A, Lozano L, Sanchez-Flores A, Rosenblueth M, et al. Genomes of *Candidatus Wolbachia bourtzisii* wDacA and *Candidatus Wolbachia pipientis* wDacB from the cochineal insect *Dactylopius coccus* (Hemiptera: Dactylopiidae). G3. 2016; **6**(10):3343-3349. DOI: 10.1534/g3.116.031237

[93] Hedges LM, Brownlie JC, O'Neill SL, Johnson KN. Wolbachia and virus protection in insects. Science. 2008;**322**: 702-702. DOI: 10.1126/science.1162418

[94] Charlat S, Hurst GDD, Merçot H. Evolutionary consequences of Wolbachia infections. Trends in Genetics. 2003;**19**:217-223

Chapter 5

The Ability of Insects to Degrade Complex Synthetic Polymers

Biswarup Mitra and Amlan Das

Abstract

Insects while feeding, encounter a wide array of hydrocarbon polymers in their diet and the digestive tracts of various insects contain microbial symbionts that aid in the degradation of these polymers. Thus the idea of insects as synthetic polymer bio-degraders was established. Soon various insect, like mealworms, flour beetles, weevils, wax moths etc. particularly from the Coleopteran and Lepidopteran orders, were identified to have remarkable abilities to consume and degrade a wide range of synthetic polymers like polyethylene, polyurethane, polypropylene, polystyrene and polyvinyl chloride into lower molecular weight, simple, and nontoxic molecules which are eventually excreted as fecula. In this review we aim at congregating the diversity of polymer degrading insect fauna and understanding the underlying mechanism in which the insect's digestive enzymes works in synergy with the gut microbiota to digest complex synthetic polymers.

Keywords: synthetic polymers, insects, gut microbiota, enzymes, degradation

1. Introduction

The vast majority of eukaryotic biodiversity in terrestrial ecosystems is represented by insects [1]. While eating, insects come into contact with a wide range of hydrocarbon polymers, and the intestinal tracts of some insects contain microbial symbionts that aid in the decomposition of these polymers. Thus, the concept of the insect as a biodegrading organism for synthetic polymers was developed. Various insects of the Coleopteran and Lepidopteran orders have been observed to have remarkable abilities to consume and degrade a wide range of synthetic polymers such as polyethylene (PE), polyurethane (PU), polypropylene (PP), polystyrene (PS), and polyvinyl chloride (PVC) into lower molecular weight, simpler, and nontoxic molecules that are eventually excreted as fecula.

Although microbial biodegradation appears sustainable, it has limits; and compared to plastic trash generation, its efficiency is modest. Furthermore, since biodegradation of a single polymer is usually a complicated process involving numerous enzymes, microbial consortia rather than a single species or strain biodegrade diverse natural and even synthesised polymers. As a result, a microbial assemblage will likely provide a more efficient biodegradation rate [2]. To overcome these limits, there was a need for a niche that would make plastic trash more accessible and bio-available to a dynamic microbial consortium. Recent research has shown that the digestive tracts of some invertebrates, notably insects, have microbial symbionts that help in the decomposition of various natural polymers that have similar structural arrangements to synthetic polymers [2, 3]. Therefore, the insect gut microbiome offered an efficient alternative for fast plastic degradation, and plastic degrading bacteria operating in concert with gut enzymes revealed increased breakdown inside the gut microbiome.

A better understanding of the function that the insect gut microbiome plays in the breakdown of plastic may be attained by actively force-feeding insects with different antibiotics and examining the variance in the molecular weight of the provided plastic feed between the insect culture with antibiotic suppressed gut microorganisms and the control insect culture without antibiotic treatment. This will allow for the acquisition of a better knowledge of the role that the insect gut microbiome plays in the degradation of plastic.

2. Insect's gut anatomy and the path to plastivory

Even though insects digest a wide range of foods, their digestive systems are largely the same. The adaptation of their diverse feeding guilds is primarily responsible for changes in their digestive tracts. The digestive tract of an insect can be structurally segmented into foregut, midgut, and hindgut. The foregut and hindgut can be divided into separate sections, each of which corresponds to a specific function. For instance, the foregut of insects is divided into pharynx and oesophagus and has a crop or diverticula for temporary food storage in addition to proventriculus for food grinding. The hindgut is separated into various regions, which include fermentation chambers and a separate rectum for retaining faeces before discharge. However, in many insects, the midgut serves as the main organ for digestion and absorption of ingested food materials [4]. Although it lacks an exoskeletal lining, the insect gut has a unique embryonic origin, having originated from endodermal cells. The peritrophic matrix serves as a protective lining for the epithelial cells lining the midgut of many insects. The peritrophic matrix divides the midgut into endo- and ectoperitrophic spaces, preventing microorganisms and abrasive food from coming into direct contact with the midgut epithelium thus preventing it from injury, pathogen infection etc. The peritrophic matrix also deactivates ingested toxins and pollutants such as pesticides and other inorganic or metal elements [5]. Furthermore, this matrix increases digestion efficiency by compartmentalising the digestion process and selectively transporting solutes and enzymes between the ectoperitrophic and endoperitrophic spaces. The peritrophic matrix further increases digestion efficiency by generating a countercurrent flow between the endo and ecto—peritrophic spaces, favouring nutrient absorption and minimising digestive enzyme loss by frass excretion [4, 6].

In this above described structure of the insect gut, resides a consortia of microorganisms which include protists, fungi, archaea, and bacteria. Fungi are common in the guts of insects that consume wood or detritus and are thought to aid digestion. Methanogenic archaea are most commonly associated with insects that feed on wood or detritus, like coleopteran beetles and isopteran termites [7, 8]. Apart from these, the most common organisms found in the almost all insect gut are a huge diversity of bacterial species. Insects that consume primarily wood as part of their diet (a behaviour known as "xylophagy") have gut microbial communities that are capable of taking part in the breakdown of cellulose [9].

Cellulose is a good source of carbon, but it appears in plant cell walls as crystalline or amorphous microfibrils, making it inaccessible to the host [10]. Here the bacteria participates to break down complex cellulose into simpler sugar residues and monosaccharides [11, 12].

The relative importance of microbial and host-derived enzymes varies as per insect species and feeding habits or diet composition. According to this theory, if insects are actively force-fed, they can degrade plastic and synthetic polymers. In general, mandibulate insects have the ability to masticate and consume plastic materials by breaking them down into smaller pieces. Even though plastic fragments are small, they have a greater surface area of contact with gut microorganisms and are therefore mixed with them. Gut microbes use the enzymes responsible for depolymerizing plastic polymers into oligomers, dimers, or monomers, and the depolymerized products are mineralised into CO_2, after which limited carbons are assimilated into biomass. Residual fragments and certain microorganisms in the gut are excreted as fecula, allowing for further degradation.

3. Synergy between insect gut microorganisms and synthetic polymers

Insect larvae owing to their capacity of consuming and absorbing synthetic polymers, especially plastic have recently opened a huge scope for researchers seeking the most efficient procedure of plastic biodegradation. Larvae of Coleopterans beetles are reported to consume and degrade plastics. *Tenebrio molitor* [13–18] and *Tenebrio obscurus* [15], the super-worm, *Zophobas asatratus* [14–16, 18], *Tribolium castaneum* [19, 20] and *Plesiophthalmus davidis* [21] etc. are few examples of members of the coleopteran order with this special ability. Besides coleopteran fauna, lepidopteran caterpillars, such as, Indian meal moth, *Plodia interpunctella* [3], the greater wax moth, *Galleria mellonella* [22–26], and the lesser wax moth, *Achroia grisella* [27] are also reported to digest synthetic polymers like polystyrene (PS), polyethylene (PE), polyvinyl chloride (PVC), and polypropylene (PP) (**Table 1**).

These insect larvae use their mandibles to consume plastics or diets that are high in plastic content. The gut symbiont and commensal microbiota of insect larvae undergo alterations when they are forcibly fed or co-fed plastic feed. In general, regardless of insect species or polymer type, consuming plastic alters the relative abundance or diversity of certain Operational Taxonomic Units (OTUs) likely Enterobacteriaceae, Enterococcaceae, and Streptococcaceae in comparison to larvae fed natural, plastic-free diets [15]. These OTUs subsequently follow a three-step process to degrade the ingested plastics: (a) microbial colonisation and biodeterioration, (b) enzymatic depolymerization (breakdown of polymer into simpler monomers) and (c) mineralisation.

The microorganisms initially colonise on the polymer either individually or in consortium (colonisation), which is assisted by various polysaccharides and/or proteins [45]. Following that, the interplaying polysaccharides and cysteine-rich proteins permeate the surface, changing the size of the polymeric pore [46]. These alterations cause biodeterioration. The durability and resilience of the polymer will decrease over time, but its surface area will expand, giving microbes a bigger surface area to adhere to. Various bacterial cells often produce an extracellular slime material that promotes adhesion and resulting in a slow positive feedback by increasing pollutant build-up, allowing for increased microbial proliferation [47]. Various bacterial cells often produce an extracellular slime material that promotes adhesion

Insect			Consumable Plastics	Insect gut microbiota	Microbe types	Interplaying enzymes	Reference
Common name	Scientific name	Order & Family					
Cigarette beetle	*Lasioderma serricorne*	Coleoptera; Ptinidae	Polyethylene (PE) Polypropylene (PP) Polyester	*Symbiotaphrina kochii*	Symbiotic yeast	Cutinase-like enzyme (CLEs)	Riudavets et al., [28]; Dowd and Shen [29]; Vega et al., [30].
Lesser grain borer	*Rhyzopertha dominica*	Coleoptera; Bostrichidae	Polyethylene (PE) Polypropylene (PP)	*Aeromonas liquifaciens*	Bacteria	Lipase Chitinase Protease	Riudavets et al., [28]; Anand and Pant [31].
Yellow Mealworm	*Tenebrio molitor*	Coleoptera; Tenebrionidae	Polystyrene (PS)	*Exiguobacterium sp* strain YT2	Bacteria	Alkaline proteases Alkali-tolerant esterase Hydrolase	Yang et al., [17]; Yang et al., [32]
			Polypropylene (PP)	*Kluyvera sp* *Citrobacter sp*	Bacteria		
			Polyethylene (PE)	*Kosakonia sp*	Bacteria		
Dark mealworm	*Tenebrio obscures*	Coleoptera; Tenebrionidae	Polystyrene (PS)	Spiroplasmataceae Enterococcaceae	Bacteria	Further research needed	Peng et al., [15]
Red flour beetle	*Tribolium castaneum*	Coleoptera; Tenebrionidae	Polystyrene (PS)	*Acinetobacter sp* AnTc-1	Bacteria		Wang et al., [19, 20]
Darkling beetle	*Plesiophthalmus davidis*	Coleoptera; Tenebrionidae	Polystyrene (PS)	*Serratia sp* strain WSW	Bacteria	Lipase Protease Chitinase	Woo et al., [21]
Lesser mealworm	*Alphitobius diaperinus*	Coleoptera; Tenebrionidae	Polystyrene (PS)	*Pseudomonas sp* *Kocuria sp* *Cronobacter sp*	Bacteria	Hydroquinone Peroxidase	Cucini et al., [33]
				Aspergillus sp *Penicillium sp* *Hyphodermella sp* *Trichoderma sp*	Fungi	Protease Cellulase Lipase	

Insect			Consumable Plastics	Insect gut microbiota	Microbe types	Interplaying enzymes	Reference
Common name	**Scientific name**	**Order & Family**					
Super worms	*Zophobas atratus*	Coleoptera; Tenebrionidae	Polystyrene (PS)	*Pseudomonas* sp strain DSM 50071 *Klebsiella pneumoniae* *Alcaligenes* sp. *Acinetobacter* sp.	Bacteria	Monooxygenase Lipase Cutinase Esterase Polyurethanase	Yang et al., [32]; Luo et al., [34]; Kim et al., [35]; Tang et al., [36]
			Polyethylene (PE)	*Citrobacter sp*			
			Polyurethane (PU)	*Mangrovibacter sp*			
Rice weevil	*Sitophilus oryzae*	Coleoptera; Curculionidae	Nylon Polyethylene(PE Polypropylene	*Bacillus subtilis* strain TLO3 *Staphylococcus sp*	Bacteria	Hydrolase Lipase	Prasad et al., [37]; Riudavets et al., [28]
Saw-toothed grain beetle	*Oryzaephilus surinamensis*	Coleoptera; Silvanidae	NylonPolyethylene	Isolation of Bacterial OTUs are yet to be done	Endosymbiotic bacteria	Further research needed	Elijah et al., [38]; Hirota et al., [39]
Greater wax moth	*Galleria mellonella*	Lepidoptera; Pyralidae	Polystyrene	*Massilia sp.* FS1903 *Acinetobacter sp* *Bacillus sp* *Serratia sp*	Bacteria	Manganese Peroxidase, Hydrogen peroxide Lac and Lignin Peroxidase (LiP) Lipase Protease Polyurethanase	Bombelli et al., [22]; Zhang et al., [40]; Jiang et al., [41]; Ren et al., [42]; Cassone et al., [23]; Lou et al., [26]
			Polyethylene	*Enterobacter* sp. D1	Bacteria		
				Aspergillus flavus	Fungi		
Indian meal moth	*Plodia interpunctella*	Lepidoptera; Pyralidae	Polyethylene	*Enterobacter asburiae* YT1 *Bacillus* sp. YP1	Bacteria	Esterase	Yang et al., [3]
Lesser waxworm	*Achroia grisella*	Lepidoptera; Pyralidae	Polyethylene	The role of the gut microbes if any on the degradation ability is yet to pondered upon.	NIL	NIL	Kundungal et al., [27]

Insect			Consumable Plastics	Insect gut microbiota	Microbe types	Interplaying enzymes	Reference
Common name	**Scientific name**	**Order & Family**					
Rice Moth	*Corcyra cephalonica*	Lepidoptera; Pyralidae	Polyethylene	*Staphylococcus saprophyticus*	Bacteria	Information is unavailable as of now	Kesti et al., [43]
Crickets	*Gryllus bimaculatus*	Orthoptera; Gryllidae	Polyester polyurethane (PUF)	*Aspergillus flavus* G10	Fungi	Hydrolytic enzymes	Khan et al., [44]

Table 1.
List of various insects and the synthetic polymers they degrade with the interplaying microbes and host enzyme.

and resulting in a slow positive feedback by increasing pollutant build up, allowing for increased microbial proliferation [47]. A number of different microbial enzymes have now initiated the enzymatic degradation process by depolymerizing and bio deteriorating the plastic polymers. Microbial enzymes (exo-enzymes) do bio-fragment synthetic polymeric structures into shorter chain oligomers, dimers, and monomers. The smaller molecules permeate and pass through the semi-permeable outer bacterial membrane (bio-assimilation) before taking up the depolymerization products (monomers) to obtain energy for cell metabolism and biomolecule production. The larvae can use the depolymerization products in the synthesis of different biomolecules.

Polymeric structures of plastics can be divided into C−C backbone and C−O backbone based on microbial breakdowns. PE, PP, PVC, and PS are examples of synthetic polymers with C−C polymeric backbones that can also be biodegraded. Microbial oxidation begins with the hydroxylation of C−C bonds and the formation of primary and secondary alcohols after the first breakdown of long-chain polymers to shorter and lower molecular weight carrying oligomers or monomers. This process is aided by the enzyme alkane hydroxylase, which does terminal and subterminal oxidation. Alcohol dehydrogenase further oxidises these alcohols, producing aldehydes and ketones. Aldehyde dehydrogenase then produces carboxylic acids, which increases the number of carbonyl-groups. The final carboxylate molecules, which are chemically identical to fatty acids, are incorporated into the oxidation pathway by microbes that provide bio-assistance for this process. In the case of PS, this generic degradation process shows only slight variation. The phenyl moieties are connected to the alternative backbone atoms of PS, which has a linear carbon backbone. Because of its unusual structure, PS biodegradation is more complicated; the organic product styrene formed after initial polymeric fragmentation is processed under the influence of numerous dioxygenase, isomerase, dehydrogenase, hydrolase, and aldolase enzymes. Easter bonds in the chemical structure of synthetic polymers with C−O backbones, such as PU and PET, increase their hydrolyzability. Polyurethane (PU) is made up of di- or poly-isocyanate and poly-ols that are linked together by carbamate (urethane) bonds [48]. Carbamate bonds connecting the crystalline stiff segments are vulnerable to attack by microorganisms. Microbial ureases, esterases, and proteases are among the enzymes that interact during PU depolymerization. During the process of PU depolymerization, ureases are responsible for breaking the urea linkage, proteases are responsible for hydrolyzing the amide and urethane linkages, and esterases are responsible for hydrolyzing the easter bonds [49] After depolymerization, the poly-ols are dehydrogenated and oxidised to produce acetyl-CoA, which is then integrated into the TCA cycle or further valorized. Terephthalic acid (TPA) and ethylene glycol (EG) are ester-bonded together to form the polymer polyethylene terephthalate (PET) [50]. The ester linkages are hydrolyzed to produce polar hydroxyl and carboxylic groups by various PET surface-modifying enzymes such as PET hydrolases after hydrolysis and depolymerization of monomeric constituents such as ethylene glycol (EG), terephthalic acid (TPA), monoethylene terephthalate (MHET), and bis-2-hydroxyethyl TPA (BHET) [51]. The enzyme MHETase is activated to further degrade the intermediate MHET and BHET into TPA and EG, which are then transported into the bacterial cell for further metabolism by dioxygenases and dehydrogenases. Finally, the final metabolites are converted into acetyl-CoA and succinyl-CoA, which enter biochemical cycles for mineralisation processes [52]. Fecula are expelled as residual and undigested particles.

4. A brief account of insects degrading synthetic polymer

4.1 Lepidoptera

The Lepidopteran insects capable of degrading synthetic polymers are discussed and detailed below. Following that, an overview of interplaying gut bacteria (**Figure 1**) that function in synergy with the host gastrointestinal enzyme is included.

4.1.1 The Indian meal-moth

Plodia interpunctella (Lepidoptera: Pyraloidea), an adult Indian mealmoth with black-tipped feet and a black to brown small-headed caterpillar, feeds on cereals, fruits, and other similar items. The waxworm caterpillars live as parasites in bee colonies and eat on pollen, cocoons, and beeswax in addition to grain mix [53]. In meal wax, at least two intestinal bacteria were found: gram-positive strain YP1, Bacillus sp., and gram-negative strain YT1, *Enterobacter asburiae* [3]. The YP1 and YT1 strains were found to have roughly 11% and 6% net loss of PE polymers, respectively [3]. During inoculation, bacteria grow on PE sheets and gain weight, resulting in the formation of a liquid suspension within about a month, and finally, the hydrophobicity and tensile strength of PE decline. PE samples become less resistant to microbial destruction as they grow less hydrophobic [54]. Both the YP1 and YT1 bacteria adhere to PE films almost immediately and form biofilms within three hours after being inoculated, indicating that they are ready for biodegradation [55, 56].

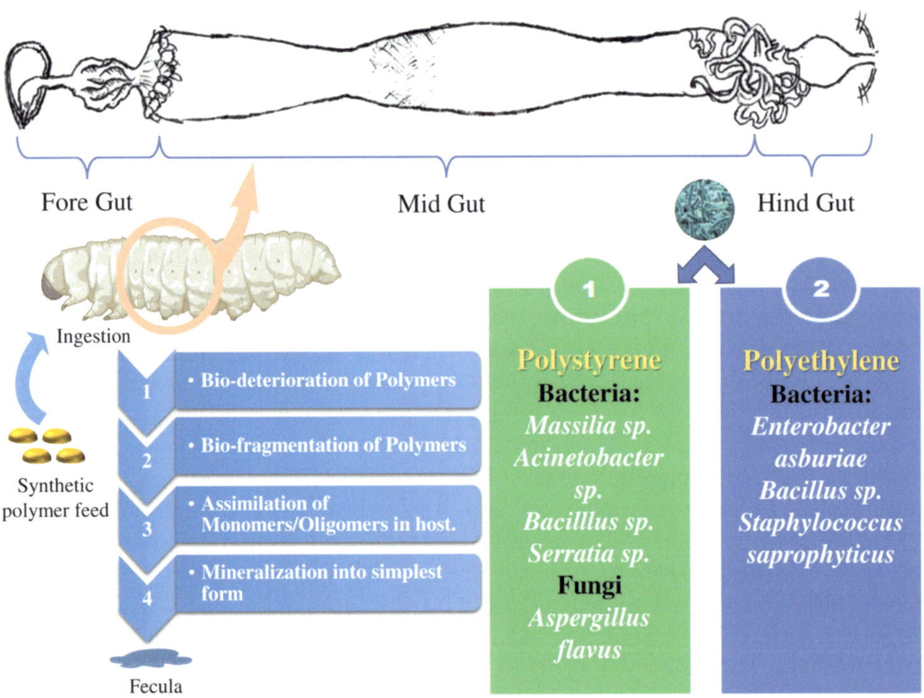

Figure 1.
Lepidopteran gut morphology and interplaying plastic degrading microbes.

A biofilm, as a non-soluble substrate, permits microorganisms to adhere to it efficiently. The presence of predominantly living bacterial strain cells on biofilm shows that PE metabolism provides these cells with the necessary nutrients [54–56]. The two bacterial strains also cause damage to the physical integrity of PE by changing surface topography, as multiple micro-pits and cavities are identified on the surface of biofilms using scanning electron microscopy (SEM) and atomic force microscopy (AFM) [3]. By increasing the quantity of carbonyl groups, the YP1 and YT1 strains elicit chemical alterations in PE [3]. The presence of the carbonyl group suggests that bacterial strains can oxidise PE materials to produce the carbonyl group, which is an important indicator of PE biodegradation. Furthermore, the weight loss of PE samples inoculated with two bacterial strains increases consistently, but the sample's molecular weight decreases. This process implies that the long-chain structure of PE is depolymerized, resulting in smaller molecular weight fragments. The chemical and physical alterations of injected PE samples show that wax worm gut bacterial strains YT1 and YP1 are capable of decomposing PE. Plastic-chewing insect larvae of the Indian meal moth, P. interpunctella, may thus represent a promising source of plastic-degrading insects.

4.1.2 The greater wax moth

The greater wax moth, *Galleria mellonella* (Lepidoptera: Pyraloidea), also known as the honeycomb moth, is a lepidopteran insect with brown-grey pigmented forewings and scaly hind wings are also reported for their plastic consuming ability. They are sexually dimorphous, 10–18 mm in length, and are distributed worldwide [57–59]. The larvae of the honeycomb moth are creamy-white, 3–30 mm in length [60], feed more intensely during earlier instars compared to later instars, and undergo eight to ten moulting stages [59]. Its larval stages are extremely damaging due to its voracious feeding habits, especially for bees and bee hives [59, 61]. The larva feeds on pollen, honey, wax, and broods and can tunnel through the comb [59, 62].

Honeycomb larva may devour PE films by generating pores and holes at a rate of more than two holes per hour per worm and can consume approximately 200 mg of PE mass in 24 h at a rate of 0.23 $mg/cm^2/h$ [22, 53]. Ethylene glycol was identified as a metabolic by-product due to PE degradation through FTIR analysis [22] or by treating caterpillars with broad-spectrum antibiotics [23]. The intestinal microbiomes of these caterpillars were found to play a distinct role in the PE degradation process [23, 40, 42]. Additionally, the larvae fed on PE showed the highest microbial abundance in their intestines, demonstrating the intestinal microbiome's favourable response to the PE diet. As a result, the presence of microbe abundance in *G. mellonella's* gut implies that the insect is benefiting metabolically from PE substrate [23]. When microbes from the intestine of *G. mellonella* were cultured in a liquid C-free medium containing PE for 60 weeks, Acinetobacter species ACT126, as well as Enterobacter sp. [42], and the fungus Aspergillus flavus [40], were discovered to be excellent candidates for contributing to the biodegradation process. Moreover, by performing Atomic Force Microscopy (AFM), an apparent change in the topography of the PE surface was observed after treating the PE with the greater wax moth. Following further microbe contact to PE films, the PE surface roughens, facilitating microbe adherence to plastic films [22].

According to the data, *G. mellonella* works naturally with flawless metabolic machinery to biodegrade lengthy hydrocarbon chains [23]. The greater wax moth feeds on beeswax in the natural, which is constituted of a highly diverse variety of lipid compounds, including alkanes, alkenes, fatty acids, and esters, with ethylene being the most common hydrocarbon bond in PE. Although more research into the molecular intricacies of wax biodegradation is needed, it appears that one of the targets of digestion is the C−C single bond of aliphatic molecules. The presence of holes in PE films exposed to waxworms and the FTIR analysis of damaged PE revealed chemical disintegration of PE, including the breakage of C−C bonds [22].

Along with PE degradation, the greater wax moth, *G. mellonella* larvae, was also reported to chew and ingest PS after analysis of their frass through GPC, FTIR, and GC-MS analysis [26]. When PS was allowed to feed on *G. mellonella* as a sole diet, the larvae could reduce PS's weight by nearly one gm in three weeks. However, co-dieting with their conventional nutritional food along with PS has resulted in increased PS degradation. The gram-positive lactic acid-producing bacteria, *Enterococcus* sp., facultatively anaerobic gram-positive bacteria, *Bacillus cereus*, and the gram-negative rod-shaped bacteria, *Serratia marcescens*, were isolated from the *G. mellonella* larval gut and were suspected of participating in PS degradation [26].

4.1.3 The lesser wax worm

An adult lesser wax worm, *Achroia grisella* (Lepidoptera: Pyraloidea), is light brown with golden highlights and black scales with long filiform antennae. Generally, it is 8–13 mm long; females are larger than males [63]. Lesser wax moths are widely distributed in tropical, subtropical, and temperate regions. The caterpillars of *A. grisella* are considered serious pests of beehives as their larvae consume bee wax [64, 65].

Like other pyraloid moths, *A. grisella* has also been reported as a PE-degrading bio-agent. They can degrade PE but less rapidly than the greater wax worm and can complete their lifecycle by consuming PE films [27]. When PE films are left in direct contact with *A. grisella* worms, the lesser wax worms, after chewing the films, make holes in them within a few days, approximately 2 ± 1 holes per worm per hour, and one individual larva can degrade nearly 2 mg of PE film daily [27].

Though the PE diet is not a good source of nutrients to grow and survive, the larvae of *A. grisella*, by consuming PE as a sole diet, live for almost one month and may develop into a second generation [27]. However, when additional nutrients were provided for them, PE degradation increased rapidly, and as a result, the survival and reproduction rate of *A. grisella* increased. In the wild, *A. grisella* larvae consume and digest beeswax, which has strong chemical bonds similar to PE. The ability of *A. grisella* larvae to digest PE plastic might be due to the presence of PE-degrading bacteria within their gut or any other unique extracellular enzymes that have not been discovered yet. The *A. grisella* caterpillar treated PE films showed an increased deviation of PE mass and decreased residual PE, suggesting most larvae consume PE either by disintegrating or assimilating the PE.

FTIR and NMR analyses of frass confirmed that the biodegradation process successfully occurs in *A. grisella* larvae [27, 66]. The presence of new carbonyl and alcoholic groups with the increase in unsaturated hydrocarbons provides evidence for the biodegradation process of PE in the lesser wax worm. However, further research must understand whether this PE biodegradation is gut-dependent or independent.

4.2 Coleoptera

Representatives of Coleoptera, capable of degrading synthetic polymers are discussed and detailed below. An overview of interplaying microbes residing in the coleopteran gut (**Figure 2**) that function in synergy with the host gastrointestinal enzyme to degrade PE, PS, and PP is included.

4.2.1 The yellow mealworm

Adult yellow-meal-worm beetles, *Tenebrio molitor* (Coleoptera: Tenebrionidae), also known as darkling beetles, are black to brown, have moniliform antennae, and complete their life cycle holo-metabolically. The larvae of yellow mealworms typically measure about 2.5 cm or more in length and have lighter body colours than adults, with long and slender structures. Generally, the mealworm feeds on stored grains, vegetation, and dead insects [67].

The larvae of mealworm beetles are capable of chewing and eating PS (Styrofoam) plastic as their sole diet [17]. Other investigations further supported this fact [13, 15, 68]. The larvae were found to degrade almost half of the consumed PS within 12–15 hours in their guts [13]. PS samples inoculated with the *Exiguobacterium* sp. bacterial strain (YT2) were found to lose more than 7% of their weight after two months of incubation [69]. The bacterial strain (YT2) was noticed to cause surface topography changes on PS materials, and as a result, the hydrophobicity of PS decreases, and carbonyl groups form. As a result, PS weight loss is due to molecular weight loss [69, 70]. It has also been opined that besides *Exiguobacterium* sp., a variety of microorganisms play an essential role in the digestion process of mealworms [71, 72].

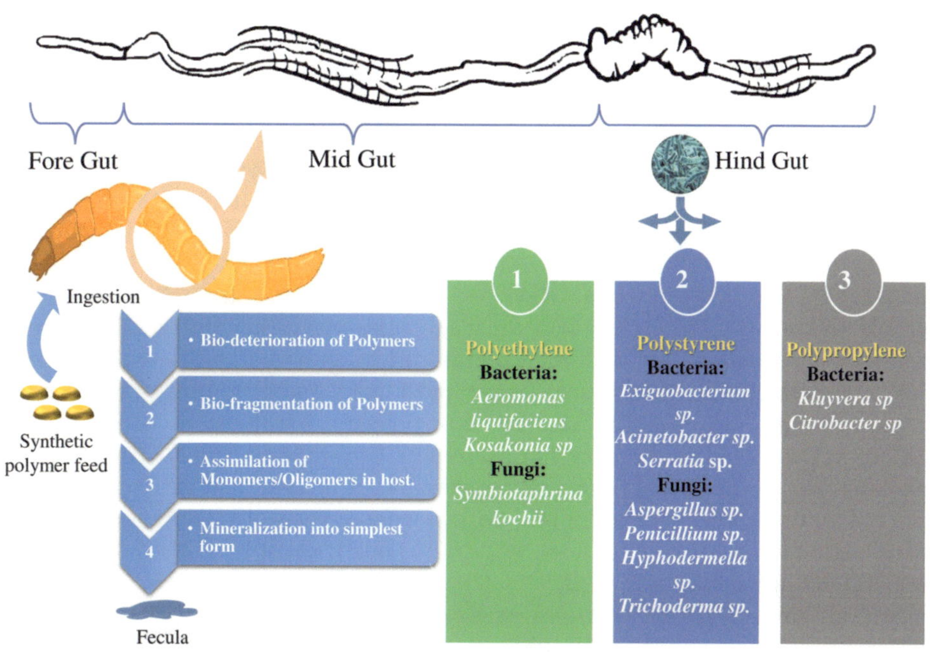

Figure 2.
Coleopteran gut morphology and interplaying plastic degrading microbes.

The information indicates that PS biodegradation and mineralisation occur within the gut of yellow mealworms [73]. During consumption, the larva generally produce hollows in Styrofoam samples, resulting in a decrease in Styrofoam mass [17] and the resultant small fragments of Styrofoam samples have an increased surface area. As a result, they were subjected to enhanced enzymatic depolymerization [17]. Another strain of mealworm (strain CA) was reported to be capable of biodegrading seven PS wastes [68]. Further investigation using mealworms from 12 different sources showed that mealworms from different regions could eat and digest PS, and those findings support the hypothesis that the capability of biodegradation of Styrofoam by mealworms is independent of their geographic origin and seems to be ubiquitous to the members of this species [68]. From this result, it could be assumed that chewing and consuming PS by yellow meal worms is their adaptive intrinsic behaviour, as they feed upon decaying forest vegetation in the wild [74]. Styrofoam-feeding mealworms had a significant survival rate, implying that Styrofoam feeding did not cause a negative effect on their survival ability [75, 76], but it was obvious that the PS degradation rate could notably be enhanced if the diet was supplemented with conventional sources of nutrition. Mealworms fed on such a diet could reproduce and enter into the second generation, which seemed to have a higher affinity for PS materials [68, 75, 76]. The temperature was also found to have corresponded with the PS degradation rate. It was found that at 250 C, the mealworm degrades PS at a significantly higher rate [68]. Moreover, PS consumption is influenced by the density of the foam materials, which is related to product hardness rather than molecular weight and thus likely to be chewed and consumed by mealworms. FTIR and NMR analysis revealed that due to cleavages at long-chains of PS molecules, they turn into low molecular weight phenyl derivative metabolites in the gut of mealworms [17, 36].

Yellow mealworms fed with PE and PS plastic each as the sole diet were found to cause mass loss of both the plastics. The yellow mealworms can degrade both PE and PS, but the degradation efficiency of PE was noticed to be much higher (48%) than PS (32%) on solo plastic diets. However, in both cases, degradation efficiency can be increased by up to 61% (for PE) and 54% (for PS) if the larvae are fed conventional food in addition to plastics [13]. The difference in mass loss of PE and PS might be due to the differences in density of the plastics, and it is presumed that less dense plastic molecules are ingested at a higher rate [68]. Among the present plastics, PE possessed a higher density than PS, which indicated that there might be other factors responsible for affecting the relative consumption rates of PE-PS plastic combinations. However, no clear evidence has yet been established to get an answer. Analysis involving HT-GPC, FTIR, and NMR studies certified that plastics could be degraded entirely and mineralised in the gut of the mealworms within a month approximately.

It was hypothesised that microbial communities significantly differed from the diets of the caterpillars or larvae. However, most microbial community members do not vary significantly in PE-fed diets among insects, but the composition is distinct in the PS-fed community. For example, *Citrobacter* sp. and *Kosakonia* sp., belonging to the family Enterobacteriaceae, were intensely associated with both plastic diets, viz., PE and PS [42] Both can use oxygen, which proves their participation in plastic degradation, as the biodegradation of both PE and PS is accelerated upon incorporating O_2 [32]. On the other hand, another two microbes, anaerobic gram-negative *Sebaldellatermitidis*, and gram-positive *Brevibacterium* sp., were uniquely associated with PE degradation [77]. Seven other microbes, viz., Listeria sp., Nitrospiradefluvii, Pedomicrobium sp., Aquihabitanssp., unclassified Xanthomonadaceae, Saprospiraceae, and Burkholeriales, were found to be associated with PS degradation in mealworm gut, further suggesting the significance of the microbial community in the plastic degradation process [78]. The information regarding the presence of various microbes in the mealworm gut suggests that mixed plastics of PS and PE could be depolymerized within the gut of the same mealworm. Therefore, the mealworm gut is not plastic-specific rather than independent in the degradation of any PE or PS plastics.

4.2.2 The dark mealworm

Adult dark meal-worm, *Tenebrio obscures* (Coleoptera: Tenebrionidae), also known as mini mealworms similar to yellow mealworms in appearance, are also known for their plastic-consuming ability. The larvae of dark meal worms are 1.5–2.5 cm in length, possess dark black rings on their abdomen, and become dark with maturity. The larvae have higher light sensitivity than yellow mealworms [15]. They usually consume seeds, vegetables, flour, and oats [68].

The larvae of dark mealworms were found to have the ability to degrade PS [15, 32], the depolymerization rate being higher than equally sized yellow mealworm larvae [15]. When dark meal worms were supplied with PS as their sole diet, mass loss of PS was found to be 55% in a month, but the amount of PS degradation was increased by 67% when the larvae were co-fed with supplementary food [15]. The investigation suggests that PS degradation ability can be achieved at a higher tempo when the insects are allowed to feed on a nutrition-rich co-diet. GPC and FTIR analysis supported that PS degradation was found to be operated by the active participation of gut bacteria residing in the dark mealworms. Before feeding PS, the gut microbiome was found to have higher diversity in *T. obscurus* than in *T. molitor* [15]. According to microbial community analysis, bacteria from the Enterococcaceae, Spiroplasmataceae, and Enterobacteriaceae families were particularly associated in their guts for PS depolymerization and degradation [15].

4.2.3 The super worm

Super worms, *Zophobas atratus* (Coleoptera: Tenebrionidae), also known as blind click-beetles, have very dark elytra on their cover, and after attaining maturity, the beetles become darker and are then called "black beetles." Superworms have mandibulate mouthparts like mealworms, which provide these species the ability to chew and eat plastic.

Super worms are also found to chew and eat Styrofoam as their sole diet [79], and when they were left on Styrofoam samples, they instantly started to ingest and penetrate through the blocks and made hollows in the blocks within an hour [68]. *Z. atratus* can consume up to 0.58 mg of Styrofoam per day, which is four times more than mealworms (0.12 mg/day/worm) [17, 32]. Interestingly, the survival rate of super worms eating Styrofoam was almost equal to that of a regular diet, which indicates that super worms can complete their lifecycle by consuming Styrofoam diets [32]. After passing through their guts, the consumed long-chain PS molecules were degraded into low molecular weight products, styrenes, which were again mineralised into CO_2 [32]. Moreover, an antibiotic suppression assay using a combination of gentamycin, rifampicin, and streptomycin indicated that repression of gut microbiota by antibiotics diminished the ability of superworms to degrade PS and, therefore, confirms that the gut microbiota plays an important role in PS degradation in superworms [32, 35, 36]. Three bacterial strains, *Aeromonas* sp. and *Klebsiella pneumonia* from Enterobacteriaceae [36] and *Pseudomonas* sp. [35] from Pseudomonadaceae have been isolated from *Z. atratus* gut and confirmed their ability to degrade PS.

4.3 Orthoptera

4.3.1 Crickets

Orthopteroid fauna like crickets, such as *Gryllus bimaculatus* are also found to consume polyurethane (PU) plastics [44]. It has been noticed that *G. bimaculatus* is capable of consuming a diet that is 63% more rich in polyurethane (PU) than its usual food. Nine distinct microbial organisms, including bacteria and fungus, were identified in their digestive tracts which might take part in PU digestion. The fungus strain *Aspergillus flavus* G10 was isolated and identified from their gut after PU-degrading activity assays. The fungus was also noted to be responsible for PU degradation [44]. However, more research needs to be done on effective insect species as well as the potent gut microbial organism that are capable to degrade PU.

4.4 Other insects

There are some other insects from coleopteran and lepidopteran order that is seen to degrade synthetic or natural polymers (**Table 1**). Insects like cigarette beetles (*Lasioderma serricorne*), lesser grain borer (*Rhyzopertha dominica*), rice weevil (*Sitophilus oryzae*), saw toothed grain beetle (*Oryzaephilus surinamensis*)—from order coleoptera are among those, that have potential in digesting polyethylene or structurally similar polymers. Other Coleopterans namely red flour beetle (*Tribolium castaneum*), darkling beetles (*Plesiopthalmus davidis*) and lesser mealworm (*Alphitobius diaperinus*) are capable of degrading polystyrene or alike polymers. Other Lepidopterans like Rice moth (*Corcyra cephalonica*) and Isopteran Termites owing to their feeding habits of complex natural polymer have immense possibilities in degrading synthetic polymers.

5. Conclusion

In recent years, there has been a significant increase in the production of plastic due to the proliferation of its usage in areas ranging from the domestic sphere to multiple business spheres. However, improper treatment and management of plastic waste disposal have led to the accumulation of this material in the environment, which poses threats to the health of living species as well as to the health of humans. The most common petroleum-based polymers, PE, PP, PS, and PVC, have been thought to be non-biodegradable for many years. However, recent studies have shown that these polymers can be degraded by the microbial communities either on their own or with the active participation of the microbial activities that are present in the larval guts of certain insects. The knowledge that is currently available about the role that insects play in the breakdown of plastic is quite restricted, and as a result, several questions on the process of plastic degradation via insects are still unclear. It has not yet been determined what the precise processes underlying the degradation process are or what the function of the enzymes should be in this process. However, the good news is that the capability of some insects to degrade compounds that are rarely biodegradable or even non-biodegradable may be employed for the practical applications for the waste management programme, which can be shown to be extremely helpful for the health of the environment.

Acknowledgements

The authors are thankful to Head, Department of Zoology, University of Calcutta for providing necessary facilities to conducting this work. The authors are also thankful to Moumita Mondal, Debjit Safui and Sumit Mondal for helping of this manuscript preparation.

Author details

Biswarup Mitra* and Amlan Das
Entomology Laboratory, Department of Zoology, University of Calcutta, Kolkata, WB, India

*Address all correspondence to: biswarupmitra1999@gmail.com

References

[1] Grimaldi D, Engel MS, Engel MS, Engel MS. Evolution of the Insects. Cambridge, UK: Cambridge University Press; May 2005:772. ISBN 0521821495

[2] Ali SS, Elsamahy T, Al-Tohamy R, Zhu D, Mahmoud Y, Koutra E, et al. Plastic wastes biodegradation: Mechanisms, challenges and future prospects. Science of the Total Environment. 2021;**780**:146590. DOI: 10.1016/j.scitotenv.2021.146590. ISSN 0048-9697

[3] Yang J, Yang Y, Wu W-M, Zhao J, Jiang L. Evidence of polyethylene biodegradation by bacterial strains from the guts of plastic-eating waxworms. Environmental Science & Technology. 2014;**48**(23):13776-13784. DOI: 10.1021/es504038a

[4] Terra WR, Ferreira C. Evolutionary trends of digestion and absorption in the major insect orders. Arthropod Structure & Development. 2020;**56**:100931

[5] Hegedus D, Erlandson M, Gillott C, Toprak U. New insights into peritrophic matrix synthesis, architecture, and function. Annual Review of Entomology. 2009;**54**(1):285-302

[6] Lehane MJ. Peritrophic matrix structure and function. Annual Review of Entomology. 1997;**42**(1):525-550

[7] Egert M, Wagner B, Lemke T, Brune A, Friedrich MW. Microbial community structure in midgut and hindgut of the humus-feeding larva of *Pachnoda ephippiata* (Coleoptera: Scarabaeidae). Applied and Environmental Microbiology. 2003;**69**(11):6659-6668

[8] Lemke T, Stingl U, Egert M, Friedrich MW, Brune A. Physicochemical conditions and microbial activities in the highly alkaline gut of the humus-feeding larva of *Pachnoda ephippiata* (Coleoptera: Scarabaeidae). Applied and Environmental Microbiology. 2003;**69**(11):6650-6658

[9] Anand R, Ramanujam J, Kulothungan S, Sharanya V, Murugalakshmi C, Bhuvaneswarik K. Plastic waste—A hazardous chemical nuisance to environment: A focus and remedy. Journal of Industrial Pollution Control. 2008;**24**(2):129-132

[10] Watanabe H, Tokuda G. Cellulolytic systems in insects. Annual Review of Entomology. 2010;**55**:609-632

[11] Hess M, Sczyrba A, Egan R, Kim TW, Chokhawala H, Schroth G, et al. Metagenomic discovery of biomass-degrading genes and genomes from cow rumen. Science. 2011;**331**(6016):463-467

[12] Warnecke F, Luginbühl P, Ivanova N, Ghassemian M, Richardson TH, Stege JT, et al. Metagenomic and functional analysis of hindgut microbiota of a wood-feeding higher termite. Nature. 2007;**450**(7169):560-565

[13] Brandon AM, Gao SH, Tian R, Ning D, Yang SS, Zhao J, et al. Biodegradation of polyethylene and plastic mixtures in mealworms (larvae of *Tenebrio molitor*) and effects on the gut microbiome. Environmental Science & Technology. 2018;**52**(11):6526-6533

[14] Brandon AM, Garcia AM, Khlystov NA, Wu WM, Criddle CS. Enhanced bioavailability and microbial biodegradation of polystyrene in an enrichment derived from the gut microbiome of *Tenebrio molitor* (mealworm larvae). Environmental Science & Technology. 2021;**55**(3):2027-2036

[15] Peng BY, Su Y, Chen Z, Chen J, Zhou X, Benbow ME, et al. Biodegradation of polystyrene by dark (*Tenebrio obscurus*) and yellow (*Tenebrio molitor*) mealworms (Coleoptera: Tenebrionidae). Environmental Science & Technology. 2019;**53**(9):5256-5265. DOI: 10.1021/acs.est.8b06963

[16] Przemieniecki SW, Kosewska A, Ciesielski S, Kosewska O. Changes in the gut microbiome and enzymatic profile of *Tenebrio molitor* larvae biodegrading cellulose, polyethylene and polystyrene waste. Environmental Pollution. 2020;**256**:113265

[17] Yang Y, Chen J, Wu WM, Zhao J, Yang J. Complete genome sequence of *Bacillus* sp. YP1, a polyethylene-degrading bacterium from waxworm's gut. Journal of Biotechnology. 2015a;**200**:77-78. DOI: 10.1016/j.jbiotec.2015.02.034

[18] Yang SS, Ding MQ, He L, Zhang CH, Li QX, Xing DF, et al. Biodegradation of polypropylene by yellow mealworms (*Tenebrio molitor*) and superworms (*Zophobas atratus*) via gut-microbe-dependent depolymerization. Science of The Total Environment. 2021b;**756**:144087. DOI: 10.1016/j.scitotenv.2020.144087

[19] Wang S, Wang L, Fan X, et al. An insight into diversity and functionalities of gut microbiota in insects. Current Microbiology. 2020a;**77**:1976-1986. DOI: 10.1007/s00284-020-02084-2

[20] Wang Z, Xin X, Shi X, Zhang Y. A polystyrene-degrading *Acinetobacter bacterium* isolated from the larvae of *Tribolium castaneum*. Science of the Total Environment. 2020b;**726**:138564. DOI: 10.1016/j.scitotenv.2020.138564

[21] Woo S, Song I, Cha HJ. Fast and facile biodegradation of polystyrene by the gut microbial flora of *Plesiophthalmus davidis* larvae. Applied and Environmental Microbiology. 2020;**86**(18):e01361-e01320. DOI: 10.1128/AEM.01361-20

[22] Bombelli P, Howe CJ, Bertocchini F. Polyethylene bio-degradation by caterpillars of the wax moth *Galleria mellonella*. Current Biology. 2017;**27**(8):R292-R293. DOI: 10.1016/j.cub.2017.02.060

[23] Cassone BJ, Grove HC, Elebute O, Villanueva SM, LeMoine CM. Role of the intestinal microbiome in low-density polyethylene degradation by caterpillar larvae of the greater wax moth, *Galleria mellonella*. Proceedings of the Royal Society B. 2020;**287**(1922):20200112. DOI: 10.1098/rspb.2020.0112

[24] Kong HG, Kim HH, Chung JH, Jun J, Lee S, Kim HM, et al. The *Galleria mellonella* hologenome supports microbiota-independent metabolism of long-chain hydrocarbon beeswax. Cell Reports. 2019;**26**(9):2451-2464. DOI: 10.1016/j.celrep.2019.02.018

[25] LeMoine CM, Grove HC, Smith CM, Cassone BJ. A very hungry caterpillar: Polyethylene metabolism and lipid homeostasis in larvae of the greater wax moth (*Galleria mellonella*). Environmental Science & Technology. 2020;**54**(22):14706-14715

[26] Lou Y, Ekaterina P, Yang SS, Lu B, Liu B, Ren N, et al. Biodegradation of polyethylene and polystyrene by greater wax moth larvae (*Galleria mellonella* L.) and the effect of co-diet supplementation on the core gut microbiome. Environmental Science & Technology. 2020;**54**(5):2821-2831

[27] Kundungal H, Gangarapu M, Sarangapani S, et al. Efficient biodegradation of polyethylene (HDPE) waste by the plastic-eating

lesser waxworm (*Achroia grisella*). Environmental Science and Pollution Research. 2019;**26**:18509-18519. DOI: 10.1007/s11356-019-05038-9

[28] Riudavets J, Salas I, Pons MJ. Damage characteristics produced by insect pests in packaging film. Journal of Stored Products Research. 2007;**43**(4):564-570. DOI: 10.1016/j.jspr.2007.03.006

[29] Dowd PF, Shen SK. The contribution of symbiotic yeast to toxin resistance of the cigarette beetle (*Lasioderma serricorne*). Entomologia Experimentalis et Applicata. 1990;**56**(3):241-248. DOI: 10.1111/j.1570-7458.1990.tb01402.x

[30] Vega FE, Dowd PF. The Role of Yeasts as Insect Endosymbionts. Insect-Fungal Associations: Ecology and Evolution. New York: Oxford University Press; 2005. pp.211-243

[31] Anand M, Pant NC. Nutrition of Rhizopertha dominica Fabricius with Reference to Possible Role of its Intracytoplasmic Microbes. Entomologists' Newsletter; 1978

[32] Yang SS, Wu WM. Biodegradation of Plastics in Tenebrio Genus (Mealworms). 2020

[33] Cucini C, Leo C, Vitale M, Frati F, Carapelli A, Nardi F. Bacterial and fungal diversity in the gut of polystyrene-fed *Alphitobius diaperinus* (Insecta: Coleoptera). Animal Gene. 2020;**17**:200109. DOI: 10.1016/j.angen.2020.200109

[34] Luo L, Wang Y, Guo H, Yang Y, Qi N, Zhao X, et al. Biodegradation of foam plastics by Zophobas atratus larvae (Coleoptera: Tenebrionidae) associated with changes of gut digestive enzymes activities and microbiome. Chemosphere. 2021. p.131006. DOI: 10.1016/j.chemosphere.2021.131006

[35] Kim HR, Lee HM, Yu HC, Jeon E, Lee S, Li J, et al. Biodegradation of polystyrene by Pseudomonas sp. isolated from the gut of superworms (larvae of *Zophobas atratus*). Environmental Science & Technology. 2020;**54**(11):6987-6996. DOI: 10.1021/acs.est.0c01495

[36] Tang ZL, Kuo TA, Liu HH. The study of the microbes degraded polystyrene. Advanced Technology and Innovation. 2017;**2**(1):13-17

[37] Prasad RK, Chatterjee S, Mazumder PB, Sharma S, Datta S, Vairale MG, et al. Study on cellulase (B-1, 4-endoglucanase) activity of gut bacteria of Sitophilus oryzae in cellulosic waste biodegradation. Bioresource Technology Reports. 2019;**7**:100274. DOI: 10.1016/j.biteb.2019.100274

[38] Elijah I, Ohimain EI, Bawo DD, Ere D, Peters EN. Biodeterioration of plastics by weevils; an environmental and stored food product perspective. International Journal of Agriculture & Environmental Science. 2015;**3**:51-55

[39] Hirota B, Okude G, Anbutsu H, Futahashi R, Moriyama M, Meng XY, et al. A novel, extremely elongated, and endocellular bacterial symbiont supports cuticle formation of a grain pest beetle. MBio. 2017;**8**(5):e01482-17. DOI: 10.1128/mBio.01482-17

[40] Zhang J, Gao D, Li Q, Zhao Y, Li L, Lin H, et al. Biodegradation of polyethylene microplastic particles by the fungus *Aspergillus flavus* from the guts of wax moth *Galleria mellonella*. Science of the Total Environment. 2020;**704**:135931. DOI: 10.1016/j.scitotenv.2019.135931

[41] Jiang S, Su T, Zhao J, Wang Z. Isolation, identification, and characterization of polystyrene-degrading bacteria from the gut of Galleria mellonella (Lepidoptera:

Pyralidae) larvae. Frontiers in Bioengineering and Biotechnology. 2021. p. 746. DOI: 10.3389/fbioe.2021.736062

[42] Ren L, Men L, Zhang Z, Guan F, Tian J, Wang B, et al. Biodegradation of Polyethylene by Enterobacter sp. D1 from the Guts of Wax Moth *Galleria mellonella*. International Journal of Environmental Research and Public Health. 2019;**16**:1941

[43] Kesti SSK, Thimmappa SCT. First report on biodegradation of low density polyethylene by rice moth larvae, Corcyra cephalonica (Stainton). The Holistic Approach to Environment. 2019;**9**(4):79-83. DOI: 10.33765/thate.9.4.2

[44] Khan S, Dong Y, Nadir S, Schaefer DA, Mortimer PE, Xu J, et al. Valorizing plastic waste by insect consumption. Circular Agricultural Systems. 2021;**1**(1):1-9

[45] Sarmah P, Rout J. Algal colonization on polythene carry bags in a domestic solid waste dumping site of Silchar town in Assam. Phykos. 2018;**48**:67-77

[46] Hakkarainen K, Ilomäki L, Lipponen L, Muukkonen H, Rahikainen M, Tuominen T, et al. Students' skills and practices of using ICT: Results of a national assessment in Finland. Computers & Education. 2000;**34**(2):103-117. DOI: 10.1016/S0360-1315(00)00007-5

[47] Zanardini E, Abbruscato P, Ghedini N, Realini M, Sorlini C. Influence of atmospheric pollutants on the biodeterioration of stone. International Biodeterioration & Biodegradation. 2000;**45**(1-2):35-42. DOI: 10.1016/S0964-8305(00)00043-3

[48] Seymour RB, Kauffman GB. Polyurethanes: A class of modern versatile materials. Journal of Chemical Education. 1992;**69**:909. DOI: 10.1021/ed069p909

[49] Ru J, Huo Y, Yang Y. Microbial degradation and valorization of plastic wastes. Frontiers in Microbiology. 2020;**11**:442. DOI: 10.3389/fmicb.2020.00442

[50] Webb HK, Arnott J, Crawford RJ, Ivanova EP. Plastic degradation and its environmental implications with special reference to poly (ethylene terephthalate). Polymers. 2013;**5**(1):1-18. DOI: 10.3390/polym5010001

[51] Kawai F, Kawabata T, Oda M. Current knowledge on enzymatic PET degradation and its possible application to waste stream management and other fields. Applied Microbiology and Biotechnology. 2019;**103**(11):4253-4268. DOI: 10.1007/s00253-019-09717-y

[52] Yoshida S, Hiraga K, Takehana T, Taniguchi I, Yamaji H, Maeda Y, et al. A bacterium that degrades and assimilates poly(ethylene terephthalate). Science. 2016;**351**:1196-1199. DOI: 10.1126/science. aad6359

[53] Khyade VB. Review on biodegradation of plastic through waxworm (order: Lepidoptera; family: Pyralidae). International Academic Journal of Economics. 2018;**5**(1):84-91

[54] Sudhakar M, Doble M, Murthy PS, Venkatesan R. Marine microbe-mediated biodegradation of low and high-density polyethylene. International Biodeterioration and Biodegradation. 2008;**61**(3):203-213

[55] Gilan I, Hadar Y, Sivan A. Colonization, biofilm formation and biodegradation of polyethylene by a strain of *Rhodococcus ruber*. Applied Microbial Biotechnology. 2004;**65**(1):97-104

[56] Sivan A, Szanto M, Pavlov V. Biofilm development of the

polyethylene-degrading bacterium *Rhododcoccus ruber*. Applied Microbial Biotechnology. 2006;**72**(20):346-352

[57] Ellis JD, Graham JR, Mortensen A. Standard methods for wax moth research. Journal of Apicultural Research. 2013;**52**:1-17

[58] Kwadha CA, Ongamo GO, Ndegwa PN, Raina SK, Fombong AT. The biology and control of the greater wax moth, *Galleria mellonella*. Insects. 2017;**8**(2):61

[59] Williams JL. Insects: Lepidoptera (moths). In: Morse R, Flottum K, editors. Honey Bee Pests, Predators, and Diseases. Medlina, OH: AI Root Company; 1997. pp. 121-141

[60] Smith T. External morphology of the larva, pupa, and adult of the wax moth, *Galleria mellonella* L. Journal of the Kansas Entomological Society. 1965;**38**:287-310

[61] Shimanuki H. Controlling the Greater Wax Moth: A Pest of Honeycombs. Washington, DC, USA: United States Department of Agriculture; 1981

[62] Turker L, Togan I, Ergezen S, Ozer M. Novel attractants of *Galleria mellonella* L. (Lepidoptera Pyralidae Galleriinae). Apidologie. 1993;**24**:425-430

[63] Cepeda-Aponte OI, Imperatriz-Fonseca VL, Velthuis HW. Lesser wax moth *Achroia grisella*: First report for stingless bees and new capture method. Journal of Apicultural Research. 2002;**41**:107-108

[64] Chandel YS, Sharma S, Verma KS. Comparative biology of the greater wax moth, *Galleria mellonella* L., and lesser wax moth, *Achoria grisella*. Forest Pest Management and Economic Zoology. 2003;**11**:69-74

[65] Mahgoub MO, Lau WH, Omar DB, El Naim AM. Life table and demographic parameters of the lesser wax moth, *Achroia grisella*, reared on natural honey bee wax. World. 2020;**8**(1):12-15

[66] Chalup A, Ayup MM, Monmany Garzia AC, Malizia A, Martin E, De Cristobal R, et al. First report of the lesser wax moth, *Achroia grisella* F. (Lepidoptera: Pyralidae) consuming polyethylene (silo-bag) in North-Western Argentina. Journal of Apicultural Research. 2018;**57**(4):569-571

[67] Zhang X, Tang H, Chen G, Qiao L, Li J, Liu B, et al. Growth performance and nutritional profile of mealworm reared on corn stover, soybean meal, and distiller's grains. European Food Research and Technology. 2019;**245**(12):2631-2640

[68] Yang SS, Wu WM, Brandon AM, Fan HQ, Receveur JP, Li Y, et al. Ubiquity of polystyrene digestion and biodegradation within yellow mealworms, larvae of *Tenebrio molitor* Linnaeus (Coleoptera: Tenebrionidae). Chemosphere. 2018;**212**:262-271

[69] Yang Y, Yang J, Wu WM, Zhao J, Song Y, Gao L, et al. Biodegradation and mineralization of polystyrene by plastic-eating mealworms: Part 1. Chemical and physical characterization and isotopic tests. Environmental Science & Technology. 2015b;**49**(20):12080-12086. DOI: 10.1021/acs.est.5b02661

[70] Wu Q, Tao H, Wong MH. Feeding and metabolism effects of three common microplastics on *Tenebrio molitor* L. Environmental Geochemistry and Health. 2019;**41**(1):17-26

[71] Genta FA, Dillon RJ, Terra WR, Ferreira C. Potential role for gut microbiota in cell wall digestion and glucoside detoxification in *Tenebrio molitor* larvae. Journal of

Insect Physiology. United Kingdom. 2006;**52**(6):593-601. DOI: 10.1016/j.jinsphys.2006.02.007

[72] Urbanek AK, Rybak J, Wrobel M, Leluk K, Mironczuk AM. A comprehensive assessment of microbiome diversity in *Tenebrio molitor* fed with polystyrene waste. Environmental Pollution. 2020:114281

[73] Kosewska O, Kosewska A, PrzemienieckiS,SienkiewiczS.Alternative ways of foamed polystyrene recycling using insects as an element of sustainable development. In: Economic Science for Rural Development Conference Proceedings. 2018;**2019**(52):45-52

[74] Howard RS. The biology of the grain beetle *Tenebrio molitor* with particular reference to its behaviour. Ecology. 1955;**36**(2):262-269

[75] Matyja K, Rybak J, Hanus-Lorenz B, Wrobel M, Rutkowski R. Effects of polystyrene diet on *Tenebrio molitor* larval growth, development and survival: Dynamic Energy Budget (DEB) model analysis. Environmental Pollution. 2020:114740

[76] Nukmal N, et al. Effect of Styrofoam waste feeds on the growth, development and fecundity of mealworms (*Tenebrio molitor*). Online Journal of Biological Sciences. Indonesia. 2018;**18**(1):24-28. DOI: 10.3844/ojbsci.2018.24.28

[77] Dwicania E, Rinanti A, Fachrul MF. Biodegradation of LLDPE plastic by mixed bacteria culture of *Pseudomonas aeruginosa* and Brevibacterium sp. Journal of Physics: Conference Series. 2019;**1402**(2):022105

[78] Lou Y, Li Y, Lu B, Liu Q, Yang SS, Liu B, et al. Response of the yellow mealworm (*Tenebrio molitor*) gut microbiome to diet shifts during polystyrene and polyethylene biodegradation. Journal of Hazardous Materials. 2021;**416**:126222

[79] Miao SJ, Zhang YL. Feeding and degradation effect on plastic of *Zhophobas morio*. Journal of Environmental Entomology. 2010;**4**

Chapter 6

Deterrents and Their Effects on the Feeding Behavior and Sensory Physiology of Insects

Vonnie D.C. Shields

Abstract

The gustatory system of insects is a prominent model in neuroscience. This important sensory system allows insects to detect, encode, and process gustatory information. This important sensory modality allows insects to perceive their environment. All animals detect and react to chemicals in their environment. Using insects as model systems allows us to obtain fundamental information regarding the processing of sensory information in the brain of the animal. Stimuli, associated with taste and smell, are responsible in insects being able to locate and select food sources, mates, and egg-laying sites. One line of research can be directed to better understanding gustatory cues in the selection of food sources by insects. Experimentally, this will involve feeding behavioral and electrophysiological testing in insects. Examining the structural organization of the gustatory organs using transmission electron and scanning electron microscopy will shed more light on the detailed structure of these taste sensory organs, the sensilla. During feeding, these taste organs sample the plant sap that contains a multitude of phytochemicals. Gustatory sensory input is encoded as patterns of nerve impulses by gustatory receptor cells which are housed in these taste sensory organs. Taste information gathered by these receptor cells will allow the insect to determine if the food is palatable or should be rejected.

Keywords: gustation, taste, glucosinolates, deterrent, feeding behavior, insect

1. Introduction

Our ability to taste is crucial for our survival and is central in our nutrition. The sense of taste determines the palatability of food and beverages. It provides early warning alerts for the detection of spoilage. Taste disorders affect the quality of life, daily living, psychological well-being and can change body weight or appetite. Having an appreciation of basic gustatory mechanisms in animals, including humans, allows us to have a better understanding and promises to contribute toward an explanation of taste disorders.

Using caterpillars as insect models allows us to increase our understanding of taste recognition and coding and to unravel some of the principles that govern food selection behavior. One feature that makes these larvae ideal candidates for such studies is their recognizable gustatory behaviors. In addition, they have relatively simple gustatory systems with relatively low numbers of sensory cells located in sensory organs (sensilla) on their mouthparts that mediate taste mechanisms. These cells are individually identifiable, show strong electrophysiological responses, and are relatively easy to access for experimental manipulation. All these features make the larval gustatory system amenable to structural, behavioral, and electrophysiological approaches, respectively.

2. Ultrastructure of main taste sensory organs: styloconic sensilla

Caterpillars bear a pair of uniporous styloconic sensory organs or sensilla (i.e., lateral and medial styloconic sensilla) on the maxillae, and more specifically, maxillary galeae. These are the main organs involved in feeding and detect plant phytochemicals by being in constant contact with plant sap during feeding. Each sensillum appears as a small cone inserted into a fibrous cuticular socket of a cylindrical projection or style of insensitive cuticle. The cone bears a terminal pore (**Figure 1**) [1]. In each styloconic sensillum, four bipolar taste neurons extend toward the tip of the cone. Receptors bound to these neurons interact with plant sap as the caterpillar is feeding. These receptors respond to salt, one or more sugars, and bitter compounds [2, 3]. One bipolar neuron, a putative mechanosensory neuron, terminates near the base of the cone and lies near the dendritic sheath [4, 5]. Here, many microtubules lie parallel to one another within a dense matrix. This location is thought to be the site of sensory transduction of mechanical stimuli [6]. Each styloconic sensillum bears a single apical or terminal pore. Gustatory sensory input gathered from the receptor cells within the sensory organs is encoded as patterns of nerve impulses which ultimately determine if relevant information is accepted or rejected in the brain of the animal (**Figure 2**).

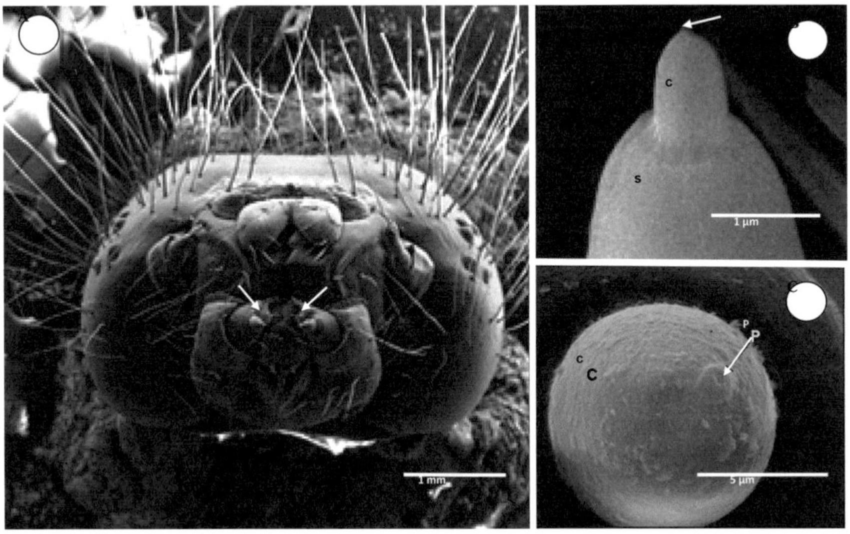

Figure 1.
A–C, Scanning electron micrographs of Lymantria dispar (L.) fifth instar larvae. The specimens were critical point dried. (A) Frontal view, whole head. The arrows denote the galeae, components of the maxillae. Bar = 1 mm. (B) Side view of a medial styloconic sensillum. The cone is inserted into the style or cylinder. The arrow shows the location of the pore at the tip of the cone. Bar = 1 μm. (C) Higher magnification view of the cone (c) from a lateral styloconic sensillum showing the pore (p with arrow) at the tip of the cone. Bar = 5 μm. This figure was adapted from [1].

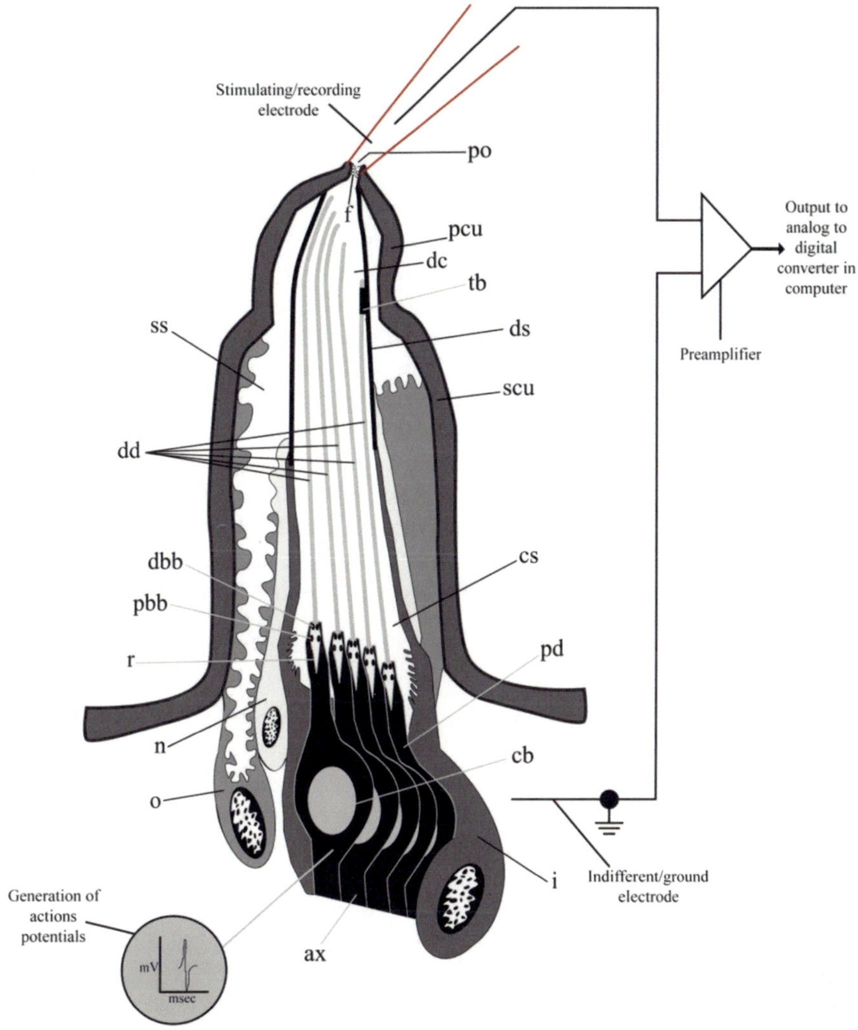

Figure 2.
Diagrammatic reconstruction of a uniporous styloconic sensillum shown in longitudinal section with five bipolar neurons innervating this sensillum: four gustatory and one mechanosensory. This illustration also shows the electrophysiological tip recording method. This method is useful for recording the neurophysiological responses from taste cells in a styloconic sensillum [7]. A taste stimulus is dissolved in an electrolyte solution (e.g., 0.1 M KCl dissolved in deionized water) contained within the stimulating or recording electrode. This electrode is placed over the tip of the pore of the sensillum. The solution diffuses through the pore. Taste compounds bind to dendritic taste receptors which transduce the quality and quantity of the stimulus into a neural code of action potentials. The indifferent or ground electrode contains a similar electrolyte solution except for the taste stimulus. Each electrode contains a silver wire. The solution and wire allow contact to be made with the internal environment of the insect (e.g., body). The excitatory responses recorded are amplified, digitized, and analyzed using a computer software program. ax, axon; cb, cell body; cs, ciliary sinus; dbb, distal basal body of proximal dendritic segment; dc, dendritic channel; dd, distal dendritic segment; ds, dendritic sheath; f, fibrils; i, inner sheath cell; n, intermediate sheath cell; o, outer sheath cell; pcu, peg cuticle; pd, proximal dendritic segment; po, terminal pore; pbb, proximal basal body of proximal dendritic segment; r, rootlets; scu, style cuticle; ss, sensillar sinus; tb, tubular body. This figure was adapted from [1].

3. Sensory responses to deterrents

At least one sensory cell is particularly sensitive to substances that cause a deterrent response in some larval Lepidoptera known as the deterrent neuron [8].

Neurophysiological responses from one or more taste cells within the sensillum can be recorded using can be acquired using an electrophysiological tip recording method [7].

Deterrent receptors in caterpillars were thought to have evolved as a proliferation of receptor types. They have an extensive action spectrum sensitive to a large variety of secondary plant compounds, ultimately resulting in appropriate behavioral outcomes that are associated with specific sensory inputs [8]. In 1992, Schoonhoven et al., hypothesized that the deterrent receptor evolved from ancestral nerve cells that retained their sensitivity to noxious plant compounds [9]. It was thought that other chemoreceptor cell types, such as sugar-sensitive cells, developed a relative insensitivity to noxious chemicals. Deterrent cells were thought to respond to compounds not previously experienced in their recent evolution [10]. If the insect transitioned to a new host-plant, a loss of sensitivity by a deterrent receptor could occur [10]. If the ingested deterrent compound, or its metabolic products, are taken up in the blood, they could potentially travel to the chemosensory cells causing a desensitization of response [11]. Over the course of evolution, the deterrent cell in the crucifer specialist, *Pieris brassicae*, became insensitive to sinigrin, probably because of this insect's very close host–plant association with the Cruciferae [12–14]. Another cell in *P. brassicae* is sensitive to glucosinolates, presumably mediates host-recognition, and likely signals acceptance rather than rejection [13].

4. Insect-plant interactions: sensory basis of feeding

The sense of taste plays a key role in the behavior of insects. Insects often rely on gustatory cues from plants to detect and find their food sources. These cues are typically nonvolatile chemicals that are either liquids or solids and can be simple or complex. They can be detected via contact chemoreceptors located on various body parts [15, 16]. Other compounds may be partially volatile [16, 17]. Examples of such compounds include DEET (N,N-diethyl-m-toluamide) [18, 19], ammonia, water, polyamines, and certain acids, pheromones, and fatty acids. Insects, in general, are selective to some extent with respect to the selection of their food choices. Monophagous insects feed on one or a few closely related plant species, whereas oligophagous insects feed on a larger number of hosts, usually confined within a certain plant family. Polyphagous insects consume many plants representing a wide taxonomic range. Insects never feed on all plant groups [20], however.

Tastants have often been grouped into taste qualities: sweet, sour, bitter, and salty [21]. Umami (savory) was added later [22]. While insects can respond to these five canonical taste qualities, taste quality perception in insects may be different than in humans. Sweet, often associated with sugars [23, 24] and sugar alcohols [23, 25–28] are attractive to Lepidoptera, as well as other insects, as are some artificial sweeteners (e.g., acesulfame K) [29]. Bitter, i.e., deterrent tastants, are represented by compounds such as caffeine, denatonium, and quinine. They may be toxic [30–33] and have diverse chemical structures [34]. Sour tastants are associated with certain acids, including acetic acid, citric acid, hydrochloric acid, and lactic acid [35–38]. Salty tastants are associated with sodium and other mineral ions, such as $NaCl$ and KCl [39–41]. Lastly, umami tastants are associated with some amino acids [24, 42–45].

Two theories exist to explain how chemical constituents of plants provide stimuli that determine food-plant preferences. Brues first suggested that insects' "botanical instinct" was based on responses to chemical and physical stimuli originating from plants [46]. Later, Dethier [47] and Fraenkel [48] stated that nutritionally unimportant "token stimuli" or attractants and repellents were primarily responsible for regulating the feeding preferences of phytophagous insects.

Many plant feeders are very specific in their food habits. They are usually restricted to a single order, family, genus, species, or even subspecies of plants. This specificity can be supported by one of two different factors, or possibly their combination: (i) insects will tend to specialize on plants that meet their dietary needs (i.e., protein, carbohydrates, fat, minerals, sterols, and vitamins) and (ii) if dietary composition is very similar for all insects, the guiding factor will be the presence or absence of additional compounds (i.e., "secondary" plant substances); the presence of nutrients would be less important. The "dual discrimination" theory proposes that insects respond to secondary compounds (token stimuli for recognition of host-plant species), as well as nutrients (for recognition of plants of the exact physiological condition nutrient content [49]). The primary "sapid" nutrients in plants act as important taste indicators of a suitable food, in addition to the recognition of secondary plant substances [50]. Host-plant selection by specialist feeders is thought to be largely influenced by the presence of token stimuli, whereas for generalist species, the presence of deterrents plays a major role [51].

Host-plant recognition and utilization, as well as avoidance or rejection of non-host plants, are generally inherited and cannot be changed by experience [52]. The primary role of secondary plant substances in insect-host-plant relationships is that they form the "fingerprint" (specific signal pattern) or biochemical profile, by which the insect identifies the plant [53]. If the plant biochemistry, as perceived by the insect, fits the expected innate image of "host-plant" to the insect, the plant will be consumed or selected as a location for egg-laying [54].

For an insect to feed, it must (a) recognize and orient to the plant; (b) begin feeding (biting or piercing); (c) maintain feeding, and (d) stop feeding, prior to dispersal. Terms applied to define classes of stimuli should encompass both physical and chemical stimuli. An "attractant" is a stimulus to which the insect responds by moving toward the food source. "Arrestants" cause the insect to stop moving toward the food source. Initiation and maintenance of feeding are separable phenomena. "Feeding incitant" is a stimulus initiating biting or piercing of the plant tissue. Once biting has started, maintenance of feeding is dependent on the presence or absence of feeding stimulants or deterrents [55]. Food selection behavior can be compared to a "key-lock" system, where the key represents a complex sensory pattern [56]. A precise behavioral response will be triggered if the pattern sufficiently corresponds with an innate standard. When the incoming sensory information differs too much from the desired pattern, the food is rejected. Host selection comprises of a series of steps (i.e., keys). Each step unlocks only one behavioral step. The lack of detail in one key will be compensated for by details in another sensory pattern (**Figure 3**) [57, 58].

The gustatory and olfactory systems of lepidopterous larvae distinguish the presence of various chemicals. Sensilla associated with these senses are located and distributed on their antennae, mouthparts, and legs. The styloconic sensilla, located on the mouthparts, are in continuous contact with plant sap during feeding. Four types of gustatory receptors have been classified into four cell types: those sensitive to nutrients, salts, phagostimulating alleochemics, and deterrents (**Figures 4** and **5**) [59, 60].

The sensitivity of chemoreceptors also vary with age, time of day, feeding history, effect of food deprivation, adaptation rate, individual insect, and temperature

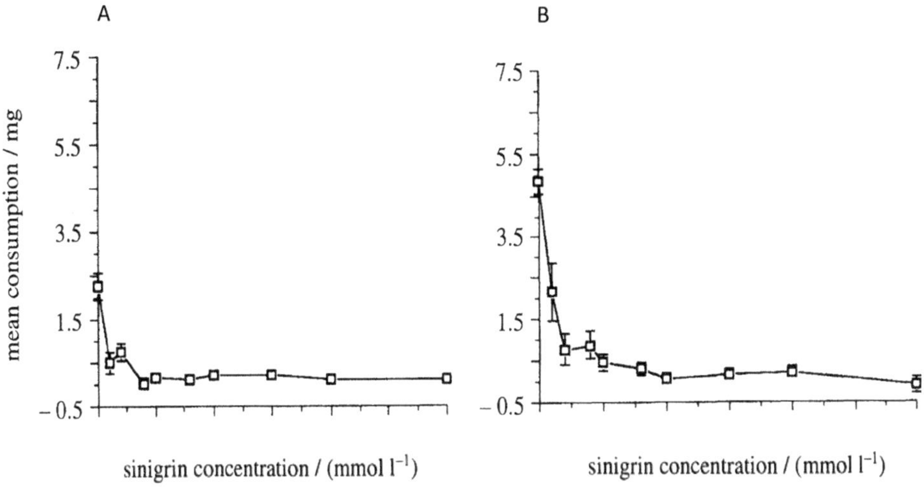

Figure 3.
A. Mean consumption by fifth instar Mamestra configurata exposed to a diet containing increasing concentrations of sinigrin. B. Mean consumption by fifth instar Trichoplusia ni exposed to a diet containing increasing concentrations of sinigrin. This figure was adapted from [58].

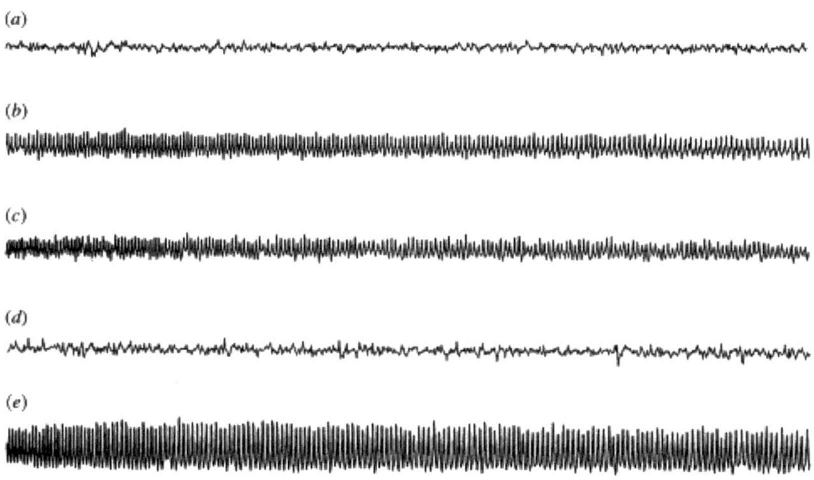

Figure 4.
Representative electrophysiological responses from the lateral and medial styloconic sensilla of fifth instar Mamestra configurata in response to: (a) 50 mM potassium chloride; (b) 2 mM sinigrin (glucosinolate); and from (c) 60 mM sucrose from lateral sensilla; (d) 50 mM potassium choride, and (e) inositol from medial sensilla. Potassium chloride (50 mM) served as the electrolyte. Note the strong firing response to inositol in (e), whereas potassium chloride evoked a very minimal response (a). This figure was adapted from [60].

[57, 59]. Städler and Hanson demonstrated in *Manduca sexta* larvae, that three of the four chemoreceptive cells, only in lateral styloconic sensilla, possessed a short-range (0–0.5 mm) olfactory capability to perceive vapors [61]. This would allow the receptors to monitor food without being in actual physical contact with it. The three mechanosensory galeal trichoid sensilla may provide information about the proximity of the food source, permitting the lateral sensillum to gauge the concentration of plant vapors accurately [61].

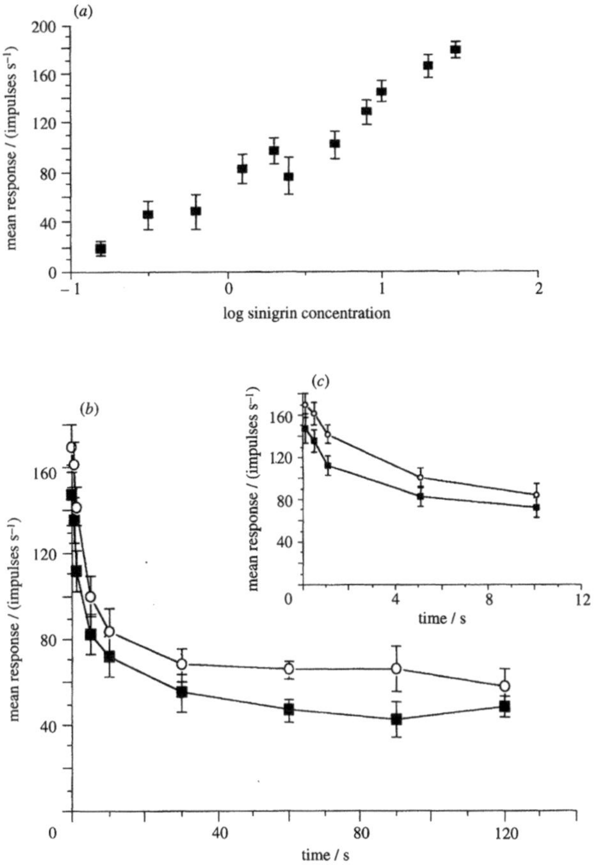

Figure 5.
(a) Dose-response curve showing the sinigrin-sensitive cell in the lateral styloconic sensillum of Mamestra configurata *when stimulated with various concentrations (mM) of sinigrin. Each point represents 10-23 larvae (cells). Error bars represent standard error of the means. (b) Adaptation curves for the sinigrin-sensitive cell in the lateral styloconic sensillum of* M. configurata *during stimulation with 8 mM (filled squares) and 20 mM (open circles) sinigrin. Each point represents means for 4-6 larvae (cells). Error bars represent standard error of the means. (c) The inset shows the first 10.1 secs. of the adaptation response for both cells. This figure was adapted from [60].*

5. Secondary plant compounds and the role of sinigrin as a feeding stimulant or deterrent

There are four major classes of secondary plant compounds: nitrogen-containing (alkaloids, amines, amino acids, cyanogenic glycosides, and glucosinolates), terpenoids (monoterpenes, sesquiterpenes, diterpenes, saponins, limonoids, cucurbitacins, cardenolides, carotenoids), phenolics (simple phenols, flavonoids including tannins, quinones), and polyacetates (polyacetylenes). These are distributed widely in vascular plants, including Solanaceae, Scrophulariaceae, Cucurbitaceae, Gymnospermae, etc. Secondary plant compounds are found in concentrations varying from, e.g., 0.0002–>40% concentration dry weight [62].

Secondary plant compounds serve as positive compound signals when an insect species becomes adapted to particular plants and uses these cues to recognize their hosts. Glucosinolates, found in the family Brassicaceae, the mustard family of flowering plants (order Brassicales), composed of 338 genera and 3700 species, appear to be limited to families of dicotyledonous angiosperms occurring in the order Capparales including the families Cruciferae, Capparaceae, Tovariaceae, Resedaceae, and Moringaceae [62]. They are present in every part of the oilseed rape plant [63] and are unlikely to serve a role in the basic metabolism of plants. Food specificity of insects is thought to be based solely on the presence or absence of these compounds [51]. Glucosinolates, or mustard oil glucosides, are derived from amino acids and contain sulfur, as well as nitrogen atoms. They can either be acyclic (e.g., sinigrin) or aromatic (e.g., sinalbin) [62].

Sinigrin (allyl- or 2-propenyl glucosinolate) is a principal crucifer token stimulus and is a widespread glucosinolate in many species of Cruciferae, as well as in other plant families. Glucosinolates are broken down by a glucosinolate-degrading enzyme (myrosinase) when plant tissues are eaten or damaged, thereby releasing toxic hydrolysis products. These products may include isothiocyanates, nitriles, thiocyanates, and oxazolidinethiones [63, 64]. Myrosinase is present in idioblasts (specialized cells in parenchymatous tissue of the green parts of crucifer plants, whereas glucosinolates are stored in vacuoles of leaf cells [64].

The role of mustard oil glucosides acting as feeding attractants, incitants, and stimulants have been studied extensively. Insects, not adapted to a particular plant species, may be repelled or deterred by the plant. For noncruciferous feeding insects, glucosinolates have been implicated as feeding deterrents. Verschaffelt demonstrated the role of a mustard oil glucoside in food-plant selection in two lepidopterous species, namely *Pieris brassicae* and *Pieris rapae* [12]. Experiments showed that these insects could be stimulated to feed on normally rejected plants by treating the plant tissue with juices extracted from crucifers. When solutions of pure sinigrin were applied to unacceptable plants, they rendered them palatable. This was further exhibited with the diamond-back moth (*Plutella maculipennis*) when a solution of either sinigrin or sinalbin (p-hydroxybenzyl glucosinolate) was applied to the leaves of nonhost plants stimulating feeding [65]. Feeding did not occur, however, when leaves were treated with the mustard oil, allyl isothiocyanate, a product of enzymatic degradation from sinigrin. The stimulatory effect of sinigrin and sinalbin was demonstrated in larval *Pieris maculipennis* feeding on a synthetic diet [66], as was the case when the addition of sinigrin strongly promoted feeding in the mustard beetle, *Phaedon cochleariae* on a synthetic diet. [67]. A similar result was observed by other researchers with synthetic diet for larval *P. brassicae* [68–71]. Isothiocyanates, such as allyl isothiocyanate, are effective in attracting larvae of the European cabbage butterly, *P. rapae* [72, 73].

Glucosinolates were found to be both deterrent and toxic to the noncrucifer-feeding lepidopterous caterpillar, *Papilio polyxenes*, which normally feeds on Umbelliferae [74]. Breakdown products may be responsible for the potency of sinigrin, as well as other glucosinolates as feeding deterrents [75]. Allyl isothiocyanate is known to be a powerful tissue irritant [64] and may be responsible for sinigrin toxicity to some species due to its release in the gut. Sinigrin may represent an innocuous form of storage in the plant, possibly as a means of avoiding autotoxicity [76]. Recently, some crucifer feeders have been shown to be deterred by some glusocinolates. Work on *Mamestra configurata* and the flea beetle, *Phyllotreta cruciferae* [77] (using the glucosinolate, sinalbin) and *Mamestra brassicae* [78], *M. configurata*, and *Trichoplusia ni* (using sinigrin [58]), clearly demonstrated deterrence in these crucifer-feeding insects.

6. Conclusions

Insects constantly monitor and respond to changes in their internal and external environments to maintain themselves under the most favorable conditions for survival. In the case of lepidopterous larvae, gustatory sensilla (i.e., lateral and medial styloconic sensilla) located on each maxilla can detect phytochemicals present in plants. They act as the first level of environmental perception and play important roles in host–plant selection, as they are in constant contact with plant sap liberated during feeding. The plant material enters each of these sensilla through an apical pore and interacts with four gustatory neurons and their receptors. Receptors bound to the dendrites transduce the chemical stimulus into a code of action potentials reflecting the quality and quantity of the complex plant chemistry. Subsequently, these nerve impulses are sent to the brain of the insect. The responses of these receptors to phytochemicals are key in determining which plants are deemed palatable and which should be rejected. Deterrent substances, such as e.g., some glucosinolates and alkaloids, are important in influencing the food selection of many insects, as they may be potentially toxic. Having a better understanding of the sensory mechanisms by which insects detect plant phytochemicals will help in finding novel biocontrol techniques against insect pests, especially highly polyphagous ones capable of defoliating forests or destroying crops.

Acknowledgements

This work was supported by NIH grants 1R15DC007609-01 and 3R15DC007609-01S1 to V.D.C.S., a grant from each of the Fisher College of Science and Mathematics, the Towson University Office of Undergraduate Research, and NIH grant DC-02751. The author gratefully acknowledges B.K. Mitchell and T. Maugel.

Author details

Vonnie D.C. Shields
Biological Sciences Department, Fisher College of Science and Mathematics, Towson University, Towson, MD, USA

*Address all correspondence to: vshields@towson.edu

References

[1] Shields VDC, Martin TL. The structure and function of taste organs in caterpillars. In: Lynch EJ, Petrov AP, editors. The Sense of Taste. Hauppauge, NY: Nova Science Publishers, Inc; 2012. pp. 147-166

[2] Shields VDC, Mitchell BK. Sinigrin as a feeding deterrent in two crucifer feeding polyphagous lepidopterous species and the effects of feeding stimulant mixtures on detterency. Philosophical Transactions of the Royal Society of London. 1995;**B347**:439-446

[3] Martin TL, Shields VDC. Detection of alkaloids and carbohydrates by taste receptor cells of the galea of gypsy moth larvae, *Lymantria dispa*r (L.) arthropod plant. Interactions. 2012;**6**:519-529

[4] Shields VDC. Ultrastructure of the uniporous sensilla on the galea of larval *Mamestra configurata* (Walker) (Lepidoptera: Noctuidae). Canadian Journal of Zoology. 1994;**72**:2016-2031. DOI: 10.1139/z94-273

[5] Shields VDC. Fine structure of the galeal styloconic sensilla of larval *Lymantria dispar* (Lepidoptera: Lymantriidae). Annals of the Entomological Society of America. 2009;**102**:1116-1125. DOI: 10.1603/008.102.0621

[6] Thurm U. Mechanoreceptos in the cuticle of the honey bee: Fine structure and stimulus mechanism. Science. 1964;**145**:1063-1065. DOI: 10.1126/science.145.3636.1063

[7] Hodgson ES, Lettvin JY, Roeder KD. Physiology of a primary chemoreceptor unit. Science. 1955;**122**:417-418. DOI: 10.1126/science.122.3166.41

[8] Dethier VG. Evolution of receptor sensitivity to secondary plant stubstances with special reference to deterrents. The American Naturalist. 1980;**115**:44-66. Available from: https://www.jstor.org/stable/2460831

[9] Schoonhoven LM, Blaney WM, Simmonds MSJ. Sensory coding of feeding deterrents in phytophagous insects. In: Bernays E, editor. Insect-Plant Interactions. Vol. IV. London: CRC Press; 1992. pp. 59-79

[10] Bernays EA, Chapman RF. The evolution of deterrent responses in plant-feeding insects. In: Chapman RF, Bernays EA, Stoffolano JG, editors. Perspectives in Chemoreception and Behavior. New York: Springer-Verlag; 1987. pp. 159-173

[11] Frazier JL. How animals perceive secondary plant compounds. In: Rosenthal GA, Berenbaum MR, editors. Herbivores: Their Interaction with Secondary Plant Metabolites. 2nd ed. Vol. vol. II. New York, NY: Academic Press Inc; 1992. pp. 89-134. DOI: 10.1016/B978-0-08-092545-5.50008-0

[12] Verschaffelt E. The cause determining the selection of food in some herbivorous insects. Proceedings of the Koninklijke Nederlandse Akademie van Wetenschappen. 1910;**13**:536-542

[13] Ma W-C. Dynamics of feeding responses in *Pieris brassicae* Linn. as a function of chemosensory input: A behavioural, ultrastructural and electrophysiological study. Meded. Landbouwhogesch. Wageningen. 1972;**72**:11. DOI: 10.1146/annurev.ento.48.091801.112629

[14] van Loon JJA. Chemoreception of phenolic acids and flavonoids in two species of *Pieris*. Journal of Comparative

Physiology A. 1990;**166**:889-899. DOI: 10.1007/BF00187336

[15] Chapman RF. Contact chemoreception in feeding by phytophagous insects. Annual Review of Entomology. 2003;**48**:455-484. DOI: 10.1146/annurev.ento.48.091801.112629

[16] Montell C. Drosophila sensory receptors—A set of molecular Swiss army knives. Genetics. 2021;**217**:1-34. DOI: 10.1093/genetics/iyaa011

[17] Delventhal R, Menuz K, Joseph R, Park J, Sun JS, Carlson JR. The taste response to ammonia in Drosophila. Scientific Reports. 2017;**7**:43754. DOI: 10.1038/srep43754

[18] Sanford J, Shields VDC, Dickens JC. Gustatory receptor neuron responds to DEET and other insect repellents in the yellow fever mosquito, *Aedes aegypti*. Naturwissenschaften. 2013;**100**:269-273. DOI: 10.1007/s00114-013-1021-x

[19] Sanford J et al. Neurophysiological and behavioral responses of gypsy moth larvae to insect repellents: DEET, IR3535, and Picaridin. PLoS One. 2014;**9**:1-7. DOI: 10.1371/journal.pone.0099924

[20] Fraenkel GS. The raison d'être of secondary plant substances. Science. 1959;**129**:1466-1470. DOI: 10.1126/science.129.3361.1466

[21] Minnich, D.E. 1921. An experimental study of the tarsal chemoreceptors of two nymphalid butterflies. The Journal of Experimental Zoology 33: 172-203. https://doi.org/10.1002/jez.1400330105.

[22] Liman ER, Zhang YV, Montell C. Peripheral coding of taste. Neuron. 2014;**81**:984-1000. DOI: 10.1016/j.neuron.2014.02.022

[23] Agnihotri AR, Roy AA, Joshi RS. Gustatory receptors in Lepidoptera: Chemosensation and beyond. Insect Molecular Biology. 2016;**25**:519-529. DOI: 10.1111/imb.12246

[24] Bestea L, Rejaud A, Sandoz JC, Carcaud J, Giurfa M, de Brito Sanchez MG. Peripheral taste detection in honey bees: What do taste receptors respond to? The European Journal of Neuroscience. 2021;**54**:4417-4444. DOI: 10.1111/ejn.15265

[25] Kikuta S, Endo. H, Tomita N, Takada T, Morita C, Asaoka K, et al. Characterization of a ligand-gated cation channel based on an inositol receptor in the silkworm, *Bombyx mori*. Insect Biochemical Molecular Biology. 2016;**74**:12-20. DOI: 10.1016/j.ibmb.2016.04.010

[26] Zhu X, Liu S, Xu C-Q, Guo K, Xu R, Qiao H-L, et al. Morphology and distribution of sensilla on the antennae and ovipositor of the hymenopteran parasitoid *Pseudotorymus jaapiellae* Yang et Chen. Microscopy Research and Technique. 2021;**84**:1873-1886. DOI: 10.1002/jemt.23744

[27] LeDue EE, Chen YC, Jung AY, Dahanukar A, Gordon MD. Pharyngeal sense organs drive robust sugar consumption in *Drosophila*. Nature Communications. 2015;**6**:6667. DOI: 10.1038/ncomms7667

[28] Takada T, Sato R, Kikuta S. A mannitol/sorbitol receptor stimulates dietary intake in *Tribolium castaneum*. PLoS One. 2017;**12**:e0186420. DOI: 10.1371/journal.pone.0186420

[29] King BH, Taylor EE, Burgess ER. Feeding response to select monosaccharides, sugar alcohols, and artificial sweeteners relative to sucrose in adult house flies, *Musca domestica*

(Diptera: Muscidae). Journal of Medical Entomology. 2019;**57**:511-518. DOI: 10.1093/jme/tjz195

[30] French A, Ali M, Agha A, Mitra A, Yanagawa A, Sellier MJ, et al. *Drosophila* bitter taste(s). Frontiers in Integrative Neuroscience. 2015;**9**:58. DOI: 10.3389/fnint.2015.00058

[31] Sollai G, Barbarossa IT, Solari P, Crnjar R. Taste discriminating capability to different bitter compounds by the larval styloconic sensilla in the insect herbivore *Papilio hospiton* (Géné). Journal of Insect Physiology. 2015;**74**:45-55. DOI: 10.1016/j.jinsphys.2015.02.004

[32] Ntie-Kang F. Mechanistic role of plant-based bitter principles and bitterness prediction for natural product studies II: Prediction tools and case studies. Physical Science Review. 2019;**4**(1126):20190007. DOI: 10.1515/psr-2019-0007

[33] Muñoz IJ, Schilman PE, Barrozo RB. Impact of alkaloids in food consumption, metabolism and survival in a blood-sucking insect. Scientific Reports. 2020;**10**:9443. DOI: 10.1038/s41598-020-65932-y

[34] Meyerhof W, Born S, Brockhoff A, Behrens M. Molecular biology of mammalian bitter taste receptors. A review. Flavour and Fragrance Journal. 2011;**26**:260-268. DOI: 10.1002/ffj.2041

[35] Crnjar R, Angioy A, Pietra P, Stoffolano JG, Liscia A, Barbarossa IT. Electrophysiological studies of gustatory and olfactory responses of the sensilla on the ovipositor of the apple maggot fly, *Rhagoletis pomo*nella Walsh. Boll. Zoology. 1989;**56**:41-46. DOI: 10.1080/11250008909355620

[36] Masala C, Loy F, Solari P, Sollai G, Muroni P, Crnjar R. Taste response 1068 profiles of the labellar chemosensilla of the medfly *Ceratitis capitata* (Diptera: Tephritidae). The Italian Journal of Zoology. 2014;**81**:32-42. DOI: 10.1080/11250003.2014.900578

[37] Mi T, Mack JO, Lee CM, Zhang YV. Molecular and cellular basis of acid taste sensation in Drosophila. Nature Communications. 2021;**12**:3730. DOI: 10.1038/s41467-021-23490-5

[38] Stanley M, Ghosh B, Weiss ZF, Christiaanse J, Gordon MD. Mechanisms of lactic acid gustatory attraction in Drosophila. Current Biology. 2021;**31**:3525-3537. DOI: 10.1016/j.cub.2021.06.005

[39] Seada MA, Ignell R, Al Assiuty AN, Anderson P. Functional characterization of the gustatory sensilla of tarsi of the female polyphagous moth *Spodoptera littoralis*. Frontiers in Physiology. 2018;**9**:1606. DOI: 10.3389/fphys.2018.01606

[40] Masagué S, Cano A, Asparch Y, Barrozo RB, Minoli S. Sensory discrimination between aversive salty and bitter tastes in an haematophagous insect. The Journal of Neuroscience. 2020;**51**:1867-1188. DOI: 10.1111/ejn.14702

[41] McDowell SAT, Stanley M, Gordon MD. A molecular mechanism for high salt taste in Drosophila. Current Biology. 2022;**32**:3070-3081. DOI: 10.1016/j.cub.2022.06.012

[42] Schoonhoven LM, van Loon JJA, Dicke M. Insect-Plant Biology. 2nd ed. Oxford: Oxford University Press; 2005

[43] Zhang YF, Huang LQ, Ge F, Wang CZ. Tarsal taste neurons of *Helicoverpa assulta* (Guenee) respond to sugars and amino acids, suggesting a role in feeding and 1449 oviposition. Journal of Insect Physiology. 2011;**57**:1332-1340

[44] Croset V, Schleyer M, Arguello JR, Gerber B, Benton R. A molecular and neuronal basis for amino acid sensing in the *Drosophila* larva. Scientific Reports. 2016;**6**:34871

[45] Aryal B, Dhakal S, Shrestha B, Lee Y. Molecular and neuronal mechanisms for amino acid taste perception in the Drosophila labellum. Current Biology. 2022;**32**:1376-1386

[46] Brues CT. The selection of food-plants by insects, with special reference to lepidopterous larvae. The American Naturalist. 1920;**54**:313-332. DOI: doi.org/10.1086/279763

[47] Dethier VG. Chemical Insect Attractants and Repellents. Philadelphia: The Blakiston Co.; 1947. 289 pp

[48] Fraenkel G. Evaluation of our thoughts on secondary plant substances. Entomologia Experimentalis et Applicata. 1969;**12**:473-486. DOI: 10.1111/j.1570-7458.1969.tb02546.x

[49] Booth JS, Booth CO. Host alternation in *Aphis fabae* Scop. I. Feeding preferences and fecundity in relation to the age and kind of leaves. Annals of Applied Biology. 1951;**38**:25-64. DOI: 10.1111/j.1744-7348.1951.tb07788.x

[50] Thorsteinson AJ. Host selection in phytophagous insects. Annual Review of Entomology. 1960;**5**:193-218. DOI: 10.1146/annurev.en.05.010160.001205

[51] Jermy T. Feeding inhibitors and food preference in chewing phytophagous insects. Entomologia Experimentalis et Applicata. 1966;**9**:1-12. DOI: 10.1111/j.1570-7458.1966.tb00973.x

[52] Jermy T. The role of experience in the host selection of phytophagous insects. In: Chapman RF, Bernays EA, Stoffolano JG, editors. Perspectives in Chemoreception and Behavior. Proceedings in Life Sciences. New York, NY: Springer; 1987. DOI: 10.1007/978-1-4612-4644-2_9

[53] Jermy T. Multiplicity of insect antifeedants in plants. In: Whitehead DL, Bowers WS, editors. Natural Products for Innovative Pest Management. Oxford: Pergamon Press; 1983. pp. 223-236

[54] Jermy T. Evolution of insect/host plant relationships. The American Naturalist. 1984;**124**:609-630. DOI: 10.1086/284302

[55] Beck SD. Resistance of plants to insects. Annals of Review. 1965;**10**:207-232. DOI: 10.1146/annurev.en.10.010165.001231

[56] Schoonhoven LM. Chemical mediators between plants and phytophagous insects. In: Nordlund DA, editor. Semiochemicals: Their Role in Pest Control. Toronto: John Wiley & Sons; 1981. pp. 31-50

[57] Schoonhoven LM. On the individuality of insect feeding behavior. In: Proceedings of the Koninklijke Nederlandse Akademie van Wetenschappen, Series C: Biological and Medical Sciences. Amsterdam, New York: North-Holland Publishing Company; 1977;**80**:341-350

[58] Shields VDC, Mitchell BK. Sinigrin as a feeding deterrent in two crucifer-feeding, polyphagous lepidopterous species and the effects of feeding stimulant mixtures on deterrency. Philosophical Transactions on Royal Society London B. 1995;**347**:439-446. DOI: 10.1098/rstb.1995.0035

[59] Schoonhoven LM. What makes a caterpillar eat? The sensory code underlying feeding behavior.

In: Chapman RF, Bernays EA, Stoffolano JG, editors. Perspectives in Chemoreception and Behavior. Proceedings in Life Sciences. New York: Springer-Verlag; 1987. pp. 69-97. DOI: 10.1007/978-1-4612-4644-2_6

[60] Shields VDC, Mitchell BK. Responses of maxillary styloconic receptors to stimulation by sinigrin, sucrose, and inositol in two crucifer-feeding, polyphagous lepidopterous species. Philosophical Transactions on Royal Society London B. 1995;**347**:447-457. DOI: 10.1098/rstb.1995.0036

[61] Städler E, Hanson FE. Olfactory capabilities of the "gustatory" chemoreceptors of the tobacco hornworm larvae. Journal of Comparative Physiology. 1975;**104**:97-102

[62] Schoonhoven LM, Van Loon JJ, Dicke M. Insect-Plant Biology. New York: Oxford University Press Inc.; 2005

[63] Fenwick GR, Heaney RK. Glucosinolates and their breakdown products in cruciferous crops, foods and feedingstuffs. Food Chemistry. 1983;**11**:249-271. DOI: 10.1016/0308-8146(83)90074-2

[64] Chew FS. Biological effects of glucosinolates. In: Biologically Active Natural Products: ACS Symposium Series. Washington D.C. 1988;**380**:155-181. DOI: 10.1021/bk-1988-0380.ch012

[65] Thorsteinson AJ. The chemotactic responses that determine host specificity in an oligophagous insect (*Plutella maculipennis* (Curt.) Lepidoptera). Canadian Journal of Zoology. 1953;**31**:52-72. DOI: 10.1139/z53-006

[66] Nayar JK, Thorsteinson AJ. Further investigations into the chemical basis of insect-host plant relationships in an oligophagous insect, *Plutella maculipennis* (Curtis) (Lepidoptera: Plutellidae). Canadian Journal of Zoology. 1963;**41**:923-929. DOI: 10.1139/z63-066

[67] Tanton MT. Agar and chemostimulant concentrations and their effect on intake of synthetic food by larvae of the mustard beetle, *Phaedon cochleariae* Fab. Entomologia Experimentalis et Applicata. 1965;**8**:74-82. DOI: 10.1111/j.1570-7458.1965.tb02344.x

[68] David WAL, Gardiner BOC. The effect of sinigrin on the feeding of *Pieris brassicae* L. larvae transferred from various diets. Entomologia Experimentalis et Applicata. 1966;**9**:95-98. DOI: 10.1111/j.1570-7458.1966.tb00980.x

[69] David WAL, Gardiner BOC. Mustard oil glucosides as feeding stimulants for *Pieris brassicae* larvae in a semi-synthetic diet. Entomologia Experimentalis et Applicata. 1966;**9**:247-255

[70] Schoonhoven LM. Chemoreception of mustard oil glucosides in larvae of *Pieris brassicae*. Proceedings of the Koninklijke Nederlandse Akademie van Wetenschappen. Series C. 1967;**70**:556-568

[71] Ma W-C. Some properties of gustation in the larva of *Pieris brassicae*. Entomologia Experimentalis et Applicata. 1969;**12**:584-590. DOI: 10.1111/j.1570-7458.1969.tb02555.x

[72] Hovanitz, W. and & Chang, V. C. S. 1963. Selection of allyl isothiocyanate by larvae of *Pieris rapae* and the inheritance of this trait. Journal of Research on the Lepidoptera, *1*(3), 169-182.

[73] Hovanitz W, Chang VCS, Honch G. The effectiveness of different isothiocyanates on attracting larvae of

Pieris rapae. Journal of Research Lepid. 1963;*4*:249-259

[74] Erickson JM, Feeny P. Sinigrin: A chemical barrier to the black swallowtail butterfly, *Papilio polyxenes*. Ecology. 1974;*55*(1):103-111. DOI: 10.2307/1934622

[75] Feeny P. Defensive ecology of the Cruciferae. Annals of Missouri Botanical Garden. 1977;**1977**:221-234. DOI: 10.2307/2395334

[76] Whittaker RH, Feeny PP. Allelochemics: Chemical interactions between species: Chemical agents are of major significance in the adaptation of species and organization of communities. Science. 1971;*171*(3973):757-770. DOI: 10.1126/science.171.3973.75

[77] Bodnaryk RP. Developmental profile of sinalbin (p-hydroxybenzyl glucosinolate) in mustard seedlings, Sinapis alba L., and its relationship to insect resistance. Journal of Chemical Ecology. 1991;**17**:1543-1556

[78] Blom F. Sensory activity and food intake: A study of input-output relationships in two phytophagous insects. Netherlands Journal of Zoology. 1978;**28**:277-340

Section 3

Vector-Borne Diseases

Chapter 7

Vector-Borne Disease and Climate Change

Amna Khan, Muhammad Yasin, Muhammad Anjum Aqueel, Muhammad Aslam Farooqi, Muhammad Irfan Akram, Hafiz Muhamad Bilal Yousuf, Muneba Noor and Aneeqa Maqsood

Abstract

Many arthropod species are vectors of numerous diseases of humans and animals, which include ticks, fleas, sandflies, mosquitoes, triatomine bugs, and blackflies. The vector transmits bacteria, viruses, and protozoa from one host to another causing various diseases, such as dengue fever, West Nile Virus, Lyme disease, and malaria. They are scold-blooded animals and very sensitive to the fluctuation in climatic factors. Changing climate influences the survival, reproduction, abundance, and spatiotemporal distribution of vectors throughout the year and the rate of development and survival of pathogens within the vector-host. Climate change is among the prime factors that influence the survival, reproduction, distribution, and density of disease vectors.

Keywords: arthropods, vector, pathogen, diseases, climate

1. Introduction

Vector-borne diseases (VBDs) continue to contribute significantly to the global burden of disease and cause epidemics that disrupt health security and cause wider socioeconomic impacts around the world [1]. A vector is an organism (most often an arthropod) that transmits an infectious pathogen from an infected human or animal host to an uninfected human. The World Health Organization (WHO) and other scientists identify the major global VBDs as malaria, dengue, chikungunya, yellow fever, Zika virus disease, lymphatic filariasis, Chagas disease, leishmaniasis, Japanese encephalitis, etc. (**Table 1**). Deadly pathogens and parasites are transmitted by arthropods [34], and the increasing global human and animal populations are threatened by such epidemics and pandemics [35]. Arthropod-borne diseases are referred to as viruses, bacteria, and protozoans transmitted either to animals or humans by blood-sucking arthropods [36]. There are over 700 known arthropod-borne viruses and at least 80 immunologically distinct types that cause diseases in humans [37].

Insect-borne diseases

Microbial class	Pathogen	Vector	Disease	Effected or host organisms	Selected references
Protozoans diseases	Plasmodium sp.	*Anopheles* sp.	Malaria	Humans and animals	[2, 3]
	Trypanosoma brucei	Glossina (tsetse fly)	Sleeping sickness, African trypanosomiasis	Humans	[4–6]
	Leishmania parasite	Phlebotomus papatasi (Sand fly)	Leishmaniasis	Animals	[4, 7]
	Trypanosoma cruzi	Triatomine bug	Chagas disease	Humans and animals	[4]
	Chikungunya virus	Aedes aegypti, Ae. Albopictus	Dengue	Humans	[8, 9]
Viral diseases	Flavivirus	*Culex sp.*	St. Louis encephalitis	Humans	[10]
	Alphavirus or Chikungunya virus	*A. aegypti*	Chikungunya fever	Humans and animals	[11, 12]
	Flavivirus Chikungunya virus	A. aegypti, Ae. Africanus	Zika disease	Humans and animals	[13]
	Chikungunya virus	*A. aegypti*	Yellow fever	Humans and animals	[11, 14]
	Flavivirus sp.	*Culex pipiens* and others	West Nile fever	Humans	[11, 14]
	Orbivirus spp.	*Culicoides* midges or *C. imicola*	Bluetongue disease	Animals	[15–17]
	Flavivirus	*Culex* spp.	Japanese encephalitis	Humans and animals	[4]
	Phlebovirus	*Aedes* spp.	Rift valley fever	Humans and animals	[18, 19]
	Yersinia pestis	*Xenopsylla cheopis*	Plague	Humans	[20, 21]

DOI: http://dx.doi.org/10.5772/intechopen.107120

Insect-borne diseases					
Bacterial diseases	*Flavivirus* sp.	*Ixodes ticks*	Tick-borne encephalitis	Animals	[4]
	Borrelia recurrentis	*Ornithodoros turicata and Pediculus humanus*	Relapsing fever	Humans and animals	[22]
	Rickettsia prowazekii or *Rickettsia typhi*	X. cheopis	Typhus	Humans	[23, 24]
	Bartonella Quintana	*P. humanus corporis*	Trench fever	Humans	[25–27]
	Wuchereria bancrofti	*Mansonia uniformis*	Malayan or lymphatic filariasis	Humans	[28]
Nematode	*Brugia malayi*	*Aedes togoi* and *C. pipiens*	Bancroftian filariasis	Humans	[29]
	Brugia malayi	*A. togoi* and *C. pipiens*	Bancroftian filariasis	Humans	[29]
Tick-borne diseases					
Protozoans disease	*Anaplasma marginale*	Ixodes sp.	Lyme disease	Humans	[11]
Bacterial disease	*Borrelia theileri*	Ixodes scapularis	*Anaplasmosis*	Animals	[30]
	Coxiella burnetii	*Rhipicephalus boophilus*	Bovine borreliosis or Tick fever	Animals	[30]
	Rickettsia rickettsia	Ixodes ricinus	Q fever	Animals	[31]
	Francisella tularensis	*Dermacentor variabilis*	Rocky Mountain spotted fever	Animals	[24, 32]
		Dermacentor andersoni, and *Haemaphysalis otophila*	Tularemia	Animals	[33]

Table 1.
Vector-borne diseases and their vector insects.

Among arthropods, insect vectors that transmit pathogens are more infectious mainly after the ingestion of the pathogen through a blood meal on an infected host, followed by the pathogens amplification/circulation in the insect's body, before the vector becomes infectious [38]. Most of the arthropod-borne viral infections are transmitted by mosquitoes. Transmission of these viruses to humans involves complex processes influenced by the mosquito and viral genetics, environmental factors, and human activities [39]. In human memory, the most devastating ABDs are not malaria or dengue or Ebola, but the plague named the Black Death, and in Europe, the Black Death is still considered the plague par excellence [40]. The major problem is caused by the invasion of exotic species of insects due to climate change [41].

The climatic factors that directly influence VBDs ecosystems are mainly temperature and rainfall [42]. Long-term variations in temperature and precipitation, climatic extremes (heatwaves, cyclones, and flash floods), quality of air, rise in low-lying coastal regions, sea level, and numerous implications on systems of food production and water supplies all have a direct impact on health. According to estimates, VBDs account for even more than 17% of the burden of communicable diseases worldwide, and more than 700,000 VBD deaths are documented each year globally [43].

2. Protozoans insect-borne disease

The health of hundreds of millions of individuals globally is seriously threatened by parasitic protozoans that are known to transmit diseases to humans. As a result of their combined infection, over a million people die each year [44, 45].

2.1 Malaria

Human malaria is caused by five species of plasmodium parasites and is transmitted by female Anopheles mosquitoes. Among vector-borne diseases, malaria is the major killer, causing an estimated 620,000 deaths in 2017 (most occurring in Africa) [46]. According to Rahmah et al. [47], malaria is currently endemic and spread by anopheline mosquitoes in far more than 80 countries with a combined population of about 3 billion people. Malaria is particularly prevalent in Sub-Saharan Africa, where even more than 85% percent of cases and 90% of deaths occur, mostly in children under the age of five. Malaria still has a staggeringly negative impact on the public's health (228 million cases globally, with 214 million, which is 93% of the total, reported in Africa alone), and recent catastrophic outbreaks have wracked many parts of the world [48, 49].

2.2 Sleeping sickness

African Trypanosomiasis, also known as sleeping sickness is caused by a microscopic parasite of the species *Trypanosoma brucei*. It is transmitted by the tsetse fly (*Glossina* spp.), which is only found in Sub-Saharan Africa [6]. The disease was perceived to be spreading across the continent from the French, Portuguese, and Belgian territorial interests in West and Central Africa towards the British and German colonies in East Africa [50]. *Trypanosoma brucei* is usually fatal within 3 years in the absence of treatment [51]. It is estimated that thousands of people are infected and millions are at risk, 1.35 million are lost due to sleeping sickness; mortality related to sleeping sickness is ranked ninth out of 25 among the human infectious and parasitic diseases in Africa [52].

2.3 Leishmaniasis

Leishmaniasis is caused by infection with Leishmania parasites, which are spread by the bite of phlebotomine sand flies. There are several different forms of leishmaniasis in people [7]. The WHO has obtained discounts on some medications to treat the disease. It is classified as a neglected tropical disease. The disease may occur in a number of other animals, including dogs and rodents. Visceral leishmaniasis is endemic in 78 countries but mainly affects economically disadvantaged populations in east Africa, Southeast Asia, and Brazil [43]. Approximately 98 nations and 4–12 million persons worldwide are now afflicted. Each year, there are about two million new cases and 20–50 thousand fatalities occurred. There are 200 million individuals who reside in regions of Asia, South and Central America, Africa, and Southern Europe where the disease is prevalent [53].

3. Viral insect-borne diseases

The viruses that cause diseases spread by insects are within a unique category and have a wide range. They are categorized as arboviruses and are made up of a diverse range of viruses that are spread by hematophagous arthropod vectors. Their primary vectors are ticks, sandflies, and mosquitoes, which are most common in tropical areas. Arboviruses like the chikungunya virus and yellow fever virus are the ones responsible for the most well-known and dangerous arthropod-borne infections. However, we must give the Flaviviridae family a unique place (Dengue, St. Louis encephalitis, Yellow fever, and West Nile encephalitis). Regarding vector-borne disease, viruses differ from one another in terms of how they behave inside of their hosts. Since humans are still unable to replicate these arboviruses to high enough titers, most human infections with them are accidental [11].

3.1 Dengue

Dengue is the most important mosquito-borne disease in the world. It is spread in tropical and subtropical regions, but the pathogen is an arbovirus, which is the general name assigned to viruses affecting mankind [9].). The viruses that cause dengue are from the Flavivirus genus, and *Aedes aegypti,* followed to a lesser extent by *Ae. albopictus*, is the primary vector. In contrast to the situation with malaria, the incidence of dengue has rapidly increased globally in recent decades. The primary distinction between dengue fever and malaria is that a virus causes both the infection and the symptoms in dengue fever [54]. Until 2005, the stated data estimated that approximately 2/5 of the world's population was susceptible to contracting dengue. However, as many instances are underreported and incorrectly classified, the number of dengue cases, like those for other insect-borne diseases, must only be used as an indication. According to reported data, 50–100 million cases of acute febrile fever occur each year, including almost 50,000 cases of severe dengue, putting more than 2 and half billion people, or more than 40% of the world's population, at risk of contracting the disease [55].

3.2 St. Louis encephalitis

St. Louis encephalitis virus (SLEV), an arthropod-borne flavivirus, can cause disease presentations ranging from mild febrile illness to severe encephalitis. It is transmitted by *Culex* sp.; birds are usually the hosts, but it can also affect humans and other hosts [56]. This disease is currently present in Canada, Mexico, and Central and South America. The clinical symptoms of this encephalitis are often not manifested, except during epidemics, wherein children and elderly are the most susceptible, with mortality 5–20% [11]. There have been cases documented across the nation, but the Mississippi Valley, the Gulf Coast, and more recently the Southwest have seen the majority of recurrent outbreaks and epidemics. SLE cases are more common in the late summer and early fall in temperate regions of the United States. In southern states, cases can happen at any time of year [57].

3.3 Chikungunya fever

Chikungunya virus (CHIKV) is an alphavirus of the Togaviridae family vectored by *Aedes* Mosquitoes [58]. CHIKV is considered as an important emerging public health problem in both tropical and temperate countries, where the distribution of the Aedes mosquito vectors continues to expand [12]. The number of cases reported to the PAHO regional office in 2016 was 349,936 suspected and 146,914 laboratories recorded incidents, which is a 50% decrease from the last previous year. Brazil reported 265,000 suspected cases, followed by Bolivia with 18,000 cases, Colombia with 19,000 cases, and Pakistan with 8387 cases [59]. The year 2016 was the first one that autochthonous transmission of chikungunya was reported in Argentina, following an outbreak of more than 1000 suspected cases [11].

3.4 Zika disease

Ae. albopictus is capable of hosting the Zika virus and is considered a potential vector for transmission among humans. This tiger mosquito is an epidemiologically important vector for the transmission of many relevant viral parasites, including yellow fever virus, dengue fever, and chikungunya fever [60, 61]. The Zika virus is the cause of the Zika disease and is a Flavivirus. The vector of the Zika virus is an *Aedes* mosquito, which can bite monkeys and humans [13]. Since 1950, ZIKA disease has been known to occur within a narrow equatorial belt from Africa to Asia. From 2007 to 2016, the virus spread eastward, across the Pacific Ocean to the Americas, leading to the 2015–s2016 Zika virus epidemic [62]. Brazil announced a connection between infection with Zika virus and microcephaly in October 2015. The Americas, Africa, and other parts of the world quickly saw outbreaks and transmission evidence. Eighty-six nations and territories in all have reported cases of Zika infection caused by mosquitoes [63]. From April 2015 to November 2016, the Zika virus, which causes Zika disease, spread throughout the Americas from Brazil. In November 2016, the WHO pronounced the outbreak to be over, but added that the virus continues to be "a highly important and long-term concern." In Brazil, the Zika virus is thought to have infected 1 and half million people, and between October 2015 and January 2016, more than 3500 newborn microcephaly cases were confirmed [64].

3.5 Yellow fever

Ae. aegypti is the most important vector of this disease. The virus spreads mostly by contact between monkeys; however, it is also capable of infecting people through a different vector. Every year, this illness results in 30,000 fatalities and 200,000 clinical cases. The majority of the cases, which affect mostly young people, are recorded from Africa, primarily the Sub-Saharan areas The disease can be fatal, acute, mild, and even unapparent [11]. Yellow fever is an acute viral hemorrhagic disease transmitted by infected mosquitoes; the infected mosquitoes of the *Aedes* spp. transmit the virus from person to person [5].

3.6 West Nile fever

Again, this disease is caused by a virus of the genus Flavivirus and the vector is an *Aedes* mosquito, aided in this case by *Culex* sp. The disease has been reported in regions of all populated continents. Birds, horses, and people can all contract West Nile fever, which has a variety of symptoms, including none at all, up until the point of death. The first virus cycle affects a number of bird species and mosquitoes, which helps the virus replicate. Secondary mammals, namely, horses and humans, are the virus's other targets [14, 65]. About 1 in 5 people who are infected develop a fever and other symptoms. About 1 out of 150 infected people develop a serious, sometimes fatal, illness [66]. Birds are exposed to the West Nile virus through the bites of infected mosquitoes. By biting infected birds, mosquitoes catch the disease. Crows and other scavengers or predatory birds like hawks and owls may contract the disease through eating sick or deceased West Nile virus-infected birds. It is possible for birds to contract the disease if they ingest infected mosquitoes.

3.7 Bluetongue disease

Bluetongue is a noncontagious, midge-borne viral disease affecting ruminants (mainly sheep and less frequently cattle, goats, antelope, deer, camel, and dromedaries). Bluetongue virus (BTV) is transmitted by Culicoides biting midges [67]. BT is a complex multi-vector, multi-host, and pathogen disease whose prevalence fluctuates, reemerging after protracted absences. The majority of transmission occurs silently in disease-resistant animals, or short devastating phenomena influencing in preference specific breeds of sheep that are economically very important for the production of high-quality wool. Most common reservoir is cattle, which can also be sub-clinically infected for a protracted infection. The name of the illness comes from its most noticeable symptom, the characteristic blue iridescent colors of the tongues of all infected animals. Although the target of this disease is limited to sheep, cattle, and some species of deer and camelids, the effects can be epidemic and devastating, causing alerts for immense economic damage. [15, 16, 68]. Changes in climate conditions in Mediterranean countries could make possible stable introductions, driving the spread of vectors and pathologic agents [17, 67, 69].

3.8 Japanese encephalitis

Japanese encephalitis (JE) is a zoonotic virus spread by vectors. In Asia, the JE virus (JEV) is the most common cause of viral encephalitis. Nearly all Asian nations, whether subtropical, temperate, or tropical, are affected by JE, which has spread to new regions through the introduction of infected vectors. The 24 nations, mostly in the WHO Western Pacific Regions and South East Asia that are now thought to be at risk of JE are home to an estimated 3 billion people. JEV repeats in an enzootic cycle in wading birds and pigs, which act as amplifying hosts and are the primary vectors of Culex mosquitoes that transmit the disease [4, 70]. It is estimated that 67,900 clinical cases of JE occur annually despite the widespread availability of the vaccine, with approximately 13,600 to 20,400 deaths. In some countries, such as Bangladesh which has no JE vaccination program, over 50% of cases occur in adults [71].

4. Bacterial insect-borne disease

4.1 Trench fever

Trench fever is a louse-borne disease caused by the gram-negative bacterium *Bartonella quintana* and observed originally in military populations during World Wars I and II. The only source of this Bartonella infection is humans. When infectious lice (Pediculus humanus corporis) excrement is rubbed into scratched skin or the conjunctiva, it can spread B. quintana to people. Trench fever is resurfacing among the US's homeless population and is endemic in México, Tunisia, Poland, Eritrea, and the former Soviet Union [26]. Trench fever had a major effect on the manpower resources of both the Allies and the Central Powers. An estimated 800,000 cases occurred among the Allies on the Western Front during World War I [72].

4.2 Relapsing fever

Spirochetes that cause tick-borne relapsing fever (TBRF) are neglected pathogens, and diagnosis of this disease is challenging because of its nonspecific manifestations [73]. Infection with *Borrelia* spp. known as relapsing fever can result in repeated episodes of fever, headache, aches in the muscles and joints, and nausea. Relapsing fever comes in three different forms: Relapsing tick-borne fever (TBRF), relapsing fever brought on by louses (LBRF), chronic *Borreliosis miyamotoi* (sometimes called hard tick relapsing fever). The western US is where TBRF occurs, and staying in shabby, rodent-infested cabins in the mountains is typically associated with it [27, 73, 74]. The mortality rate is 1% with treatment and 30–70% without treatment [75]. In Texas, TBRF is frequently linked to cave exposures. LBRF is transmitted by the human body louse and usually occurs in refugee settings in developing parts of the world [76].

4.3 Plague

The gram-negative coccobacillus Yersinia pestis, which is 2 μm long, is what causes the plague. Pneumonic, systemic infection, and bubonic plague are the three primary illness manifestations caused by the infection in humans. The final one stands out and is well-known [21]. It spreads in the typical arthropod-borne illness triangle of pathogen, vector, and victim. The vector in this instance is the Oriental flea (Xenopsylla cheopis), which feeds on an infected animal and regurgitates blood containing parasitic cells [20, 77].

5. Tick-borne disease

Ticks are ectoparasites that are common throughout the world, and their eco epidemiology is directly tied to environmental factors. They are required hematophagous ectoparasites that act as reservoirs or carriers for dangerous bacteria, viruses, protozoa, rickettsia, and another fungus while they feed on their hosts. The next vector group that mostly spreads infections to humans and primarily affects animals in the disease transmission process is the tick [78].

5.1 Lyme disease

Lyme disease is the most common disease spread by ticks in the Northern Hemisphere, with an estimated 300,000 infected people a year in the USA alone and 65,000 people a year in Europe. The infection is caused by a bacterium of the genus *Borrelia,* spread by ticks, but it is transmitted to humans by the bites of infected ticks of the genus *Ixodes*. The disease is more common in the spring and early summer [11].

In Europe, *Ixodes ricinus* spreads a large number of tick-borne infections, such as *Lyme borreliosis* and tick-borne encephalitis. Potential habitat expansion in northern Europe was forecast by models, along with warmer winter weather conditions including temperature rise. These circumstances might make it easier for more ticks to survive the winter and raise the risk of tick bites [79].

The higher temperature was found to be the most important determinant of environmental suitability for the establishment of the Lyme disease Ixodes tick vector in southern Canada, where it has been spreading [80]. Milder and shorter winters in Quebec, Canada are associated with the northern spread of the white-footed mouse, the primary reservoir host for the Lyme disease pathogen *Borrelia burgdorferi* [81].

5.2 Tularemia

Tularemia is a highly contagious disease occurring principally in wild animals, but it may transmit to farm animals, causing septicaemia and high mortality. *Francisella tularensis* is the causative organism [33]. *F. tularensis* has a wide host range and is recorded in over 100 species of bird and wild and domestic animals [82]. The mortality rate is 50%, especially in young animals. Transmission occurs chiefly by the bites of the wood tick, *Dermacentor andersoni*, and from *Haemaphysalis otophila*. Tularemia is primarily restricted in its occurrence to countries in the Northern Hemisphere. In North America, the disease is most prevalent in farm animals in the North Western states of the USA and the adjoining areas of Canada [83].

6. Importance of climate change on vector-borne disease (VBD)

There is no doubt that a suitable climate is necessary for the persistence or emergence of a vector-borne disease [84]. Since a warmer temperature and shifting rainfall patterns may produce favorable habitats for viruses and climate-sensitive vectors (such as mosquitoes and ticks), climate change has an impact on how VBDs are transmitted. Climate change is recognized as a significant driver affecting the epidemiology and dissemination of VBDs on various time scales, albeit the implications are exacerbated by nonlinear feedback, intrinsic in the dynamics of infections [85].

6.1 How does climate affect VBD

Through a variety of mechanisms, including direct impacts on the pathogens, the vector, nonhuman hosts, and humans, climate can influence the dynamics of transmission, geographic spread, and reemergence of vector-borne diseases. Climate change can change entire ecosystem habitats, including urban habitats, which can either benefit or harm vectors or nonhuman hosts. This is in addition to getting direct effects on specific species [84].

6.2 Warm climate and changing rainfall

Since arthropods and many other vectors are ectotherms, it is anticipated that with rising temperatures, both the rate of pathogen development and the length of the transmission season in some endemic locations, as well as the abundance, survival, and feeding activity of vectors, will rise. (The other dengue vector, *Aedes albopictus*, has already shown widespread growth into further temperate regions; it is unknown what impact climate change will play in this.) If endemic areas get so hot that vector survivability or feeding is hindered, there is also a chance that the prevalence of dengue or other vector-borne diseases would diminish. But the effects of excessive heat would still be very bad in these regions. Rainfall and vector abundance have a complicated and context-specific relationship [86].

6.3 Climate is an effective determinant of VBD

Vector-borne diseases are among the most well studied diseases associated with climate change, owing to their large disease burden, widespread occurrence, and high sensitivity to climatic factors [1]. The natural environment and human systems are also affected by climate and weather conditions in a variety of more indirect ways. For instance, a drought may have an impact on water storage, soil use and irrigation techniques, and population movement, which in turn may have an impact on vector ecology and human infection exposure [87]. The most direct links exist between temperature and the rates of biting, reproduction, and survival of vectors as well as the development and survival of the infections they carry. Additionally, precipitation has a significant impact on diseases that are spread by vectors with aquatic developmental stages, such as mosquitoes, as well as diseases that are spread by vectors without any such stages, including ticks or sandflies [88]. The impact of climate on the transmission of disease may be obscured, for instance, when the pathogens take a long time to develop in the host or when the vectors are relatively protected from weather and climate because they spend all of their time indoors (such as triatomine bugs, which transmit Chagas disease) (such as the nematode worms, which cause filariasis). Climate change can have quite varied effects on diseases even whose transmission cycles appear to be identical. Consider the differences between malaria and dengue, two of the most significant and thoroughly researched vector-borne diseases, both anthroponoses spread by mosquitoes [55].

7. Future impact of climate change

This chapter focuses on major changes in vector and pathogen dissemination that have been observed in recent years in temperate, Arctic, Peri-arctic as well as tropical highland regions. These changes have been predicted by scientists all over the world. In addition, if we do not mitigate and respond to climate change, further changes are probably coming. The movement of people, animals, and goods; existing control measures; the availability of efficient medications; the standard of health services; human behavior; and political stability and conflicts are just a few of the important factors that influence the spread and intensity of human diseases.

Author details

Amna Khan[1], Muhammad Yasin[2*], Muhammad Anjum Aqueel[2], Muhammad Aslam Farooqi[2], Muhammad Irfan Akram[2], Hafiz Muhamad Bilal Yousuf[2], Muneba Noor[2] and Aneeqa Maqsood[2]

1 Faculty of Agricultural Science and Technology, Department of Agricultural Entomology, Bahauddin Zakariya University, Multan, Pakistan

2 Faculty of Agriculture and Environment, Department of Entomology, The Islamia University of Bahawalpur, Bahawalpur, Pakistan

*Address all correspondence to: yasin_1876@yahoo.com

References

[1] Campbell-Lendrum D, Manga L, Bagayoko M, Sommerfeld J. Climate change and vector-borne diseases: What are the implications for public health research and policy? Philosophy Transaction of the Royal Society B. 2015;**370**:20130552

[2] Sallares R, Bouwman A, Anderung C. The spread of malaria to southern Europe in antiquity: New approaches to old problems. Medical History. 2004;**48**(3):311-328

[3] Otto TD, Gilabert A, Crellen T, Böhme U, Arnathau C, Sanders M, et al. Genomes of all known members of a Plasmodium subgenus reveal paths to virulent human malaria. Nature Microbiology. 2018;**3**(6):687-697

[4] WHO and UNICEF. 2017. Global vector control response 2017-2030

[5] WHO. 2022. Fact sheet Yellow fever. Available at: https://www.who.int/news-room/fact-sheets/detail/yellow-fever. [Accessed on April 12, 2022]

[6] Center of disease control (CDC). 2022a. Parasite-African trypanosomiasis (also known as sleeping sickness). Available at: https://www.cdc.gov/parasites/sleepingsickness/index.html. [Assessed on April 16, 2022]

[7] Center of disease control (CDC). 2022b. Parasite-Leishmaniasis. Available at: https://www.cdc.gov/parasites/leishmaniasis/index.html. [Assessed on April 20, 2022]

[8] Roiz D. Climatic factors driving invasion of the tiger mosquito (*Aedes albopictus*) into new areas of Trentino, northern Italy. PLoS One. 2001;**6**: e14800

[9] Peng HJ, Lai HB, Zhang QL, Xu BY, Zhang H, Liu WH, et al. A local outbreak of dengue caused by an imported case in Dongguan China. BMC Public Health. 2012;**12**(1):1-8

[10] Reisen WK. Epidemiology of St. Louis encephalitis virus. Advance Virus Research. 2003;**61**:139-183

[11] Marcello N. Three Scenarios in insect-insect borne diseases. In: Nicoletti M, editor. Insect-insect Borne Diseases in the 21st Century. Academic Press; 2020. pp. 99-251

[12] Dubrulle M, Mousson L, Moutailler S, Vazeille M, Failloux AB. Chikungunya virus and Aedes mosquitoes: Saliva is infectious as soon as two days after oral infection. PLoS One. 2009;**4**(6):e5895

[13] Benelli G. Spread of Zika virus: The key role of mosquito vector control. Asian Pacific Journal of Tropical Medicine. 2016;**6**(6):468-471

[14] Mackenzie JS, Gubler DJ, Petersen LR. Emergence favivirus; the spread and resurgence of Japanese encephalitis, West Nile fever and Dengue. Nature Medicine. 2004;**10**(12):98-109

[15] Bowne JG. Bluetongue disease. Advances in Veterinary Science and Comparative Medicine. 1971;**15**:1-46

[16] Saegerman S, Berkvens D, Mellor PS. Bluetongue epidemiology in the European Union. Emerging Infectious Diseases. 2008;**14**:539-544

[17] Pioz M, Guis H, Crespin L, Gay E, Calavas D, Durand B, et al. 2012. Why did bluetongue spread the way it did? Environmental factors influencing the

velocity of bluetongue virus serotype 8 epizootic wave in France

[18] Tshilenge G. Immunogeno: Protective mechanism for Rift valley fever in the Democratic Republic of Congo. Journal of Veternary Research. 2012;**79**:1-35

[19] Georges TM, Justin M, Victor M, Marie KJ, Mark R, Léopold MMK. Seroprevalence and virus activity of Rift valley fever in cattle in eastern region of Democratic Republic of the Congo. Journal of Veternary Medicine. 2018;**2018**:4956378

[20] Carter R, Mendis KN. Evolutionary and historical aspects of the burden of malaria. Clinical Microbiology Reviews. 2002;**15**(4):564-594

[21] Shrewsbury JFD. A history of Bubonic plague in the British Isles. Cambridge University press, UK; 2005. p. 3

[22] Reck J, Marks FS, Guimarães JA, Termignoni C, Martins JR. Epidemiology of *Ornithodoros brasiliensis* (mouro tick) in the southern Brazilian highlands and the description of human and animal retrospective cases of tick parasitism. Ticks and Tick Borne Diseases. 2013;**4**:101-109

[23] Raoult D, Ndihokubwayo JB, Tissot-Dupont H, Roux V, Faugere B, Abegbinni R, et al. Outbreak of epidemic typhus associated with trench fever in Burundi. The Lancet. 1998;**352**(9125):353-358

[24] Walker DH, Raoult D. *Rickettsia rickettsii* and other spotted fever group Rickettsiae (Rocky Mountain spotted fever and other spotted fevers). In: Mandell GL, Bennet JE, Doalin R, editors. Principles and Practice of Infectious Diseases. Philadelphia: Churchill Livingstone; 2000. pp. 2035-2042

[25] Ohl ME, Spach DH. *Bartonella quintana* and urban trench fever. Clinical Infectious Diseases. 2000;**31**(1):131-135

[26] Anstead Gregory M. The centenary of the discovery of trench fever, an emerging infectious disease of World War 1. The Lancet Infectious Diseases. 2016;**16**(8):164-172

[27] Larry, M. 2022. Spirochetes relapsing fever. Available at: https://www.msdmanuals.com/professional/infectious-diseases/spirochetes/relapsing-fever. [Assessed on April 21, 2022]

[28] Ughasi J, Bekard HE, Coulibaly M, Adabie-Gomez D, Gyapong J, Appawu M, et al. *Mansonia africana* and *Mansonia uniformis* are vectors in the transmission of *Wuchereria bancrofti* lymphatic filariasis in Ghana. Parasite and Vectors. 2012;**5**:89

[29] Wada Y. Vector mosquitoes of filariasis in Japan. Tropical Medicine and Health. 2011;**39**:39-45

[30] Barker SC, Walker AR. Ticks of Australia. The species that infest domestic animals and humans. Zootaxa. 2014;**3816**(1):1-144

[31] Raoult D, Marrie TJ, Mege JL. Natural history and pathophysiology of Q fever. Lancet Infectious Diseases. 2005;**5**:219-226

[32] McDonald GA, Anacker RL, Mann RE, Milch LJ. Protection of guinea pigs from experimental Rocky Mountain spotted fever with a cloned antigen of *Rickettsia rickettsii*. Journal of Infectious Diseases. 1988;**158**:228-231

[33] Kocan KM, DE LA Fuente J, Guglielmone AA, Meléndez RD. Antigens and alternatives for control of *Anaplasma marginale* infection in cattle. Clinical Microbiology Reviews. 2003;**16**:698-712

[34] Mehlhorn H. Encyclopedia of Parasitology. Vol. 1. Düsseldorf, Germany: Institut für Zoomorphologie, Zellbiologie und Parasitologie; 2016

[35] Mehlhorn H, Al-Rasheid KAS, Al-Quraishy S, Abdel-Ghaffar F. Research and increase of expertise in arachno-entomology are urgently needed. Parasitology Research. 2012;**110**:259-265

[36] Gubler DJ. The global emergence/resurgence of arboviral diseases as public health problems. Archives of Medical Research. 2002;**33**:330-342

[37] Smith KM, Nanda K, Spears CJ, Ribeiro M, Vancini R, Piper A, et al. Structural mutants of dengue virus 2 transmembrane domains exhibit host-range phenotype. Virology Journal. 2011;**8**(1):1-15

[38] Wilson AJ, Morgan ER, Booth M, Norman R, Perkins SE, Hauffe HC, et al. What is a vector? Philosophical Transactions of the Royal Society B. 2017;**372**(1719)

[39] Reiter P. Climate change and mosquito-borne disease. Environmental Health Perspective. 2001;**109**:141-161

[40] Benedictow OJ. The Black Death 1346-1353: The Complete History. New York: Boydell Press; 2004

[41] Yasin M, Khan A, Wakil W, Qayyum MA, Rubab S. Microbial management of ornamental plants/palm common pests. In: Kumar A, editor. Microbial Biocontrol: Food Security and Post-Harvest Management. Cham: Springer; 2022

[42] Fouque F, Reeder JC. Impact of past and on-going changes on climate and weather on vector-borne diseases transmission: A look at the evidence. Infectious Diseases of Poverty. 2019;**8**(51):9-20

[43] WHO. Global leishmaniasis surveillance, 2017–2018, and first report on five additional indicators. Weekly Epidemiology Record. 2020a;**25**:265-280

[44] Andrews KT, Fisher G, Skinner-Adams TS. Drug repurposing and human parasitic protozoan diseases. International Journal of Parasitology, Drugs Drug Resistance. 2014;**4**:95-111

[45] Lozano R, Naghavi M, Foreman K, Lim S, Shibuya K, Aboyans V, et al. Global and regional mortality from 235 causes of death for 20 age groups in 1990 and 2010: A systematic analysis for the global burden of disease study 2010. The Lancet. 2012;**380**(9859):2095-2128

[46] Global, regional, and national age-sex-specific mortality for 282 causes of death in 195 countries and territories, 1980-2017: a systematic analysis for the Global Burden of Disease Study 2017. GBD, Causes of death collaborators. Lancet. 2018;**392**:1736-1788

[47] Rahmah Z, Sasmito SD, Siswanto B, Sardjono TW, Fitri LE. Malaria. Malaysian Journal of Medical Science. 2015;**22**:25-32

[48] Amato R, Pearson RD, Almagro-Garcia J, Amaratunga C, Lim P, Suon S, et al. Origins of the current outbreak of multidrug-resistant malaria in South East Asia: A retrospective genetic study. The Lancet Infectious Diseases. 2018;**18**(3):337-345

[49] Lok P, Dijk S. Malaria outbreak in Burundi reaches epidemic levels with 5.7 million infected this year. BMJ. 2019;**366**

[50] Welburn SCI, Maudlin I, Simarro PP. Controlling sleeping sickness - A review. Parasitology. 2009;**136**:1943-1949

[51] Checchi F, Filipe JA, Haydon DT, Chandramohan D, Chappuis F. Estimates of the duration of the early and late stage of gambiense sleeping sickness. BMC Infectious Diseases. 2008;**8**:16

[52] Fevre EM, Wissmann BV, Welburn SC, Lutumba P. The burden of human African trypanosomiasis. PLoS Neglected Tropical Disease. 2008;**2**:e333

[53] Vos T, Allen C, Arora M, Barber RM, Bhutta ZA, Brown A, et al. (GBD 2015 Disease Injury Incidence Prevalence Collaborators). Global, regional, and national incidence, prevalence, and years lived with disability for 310 diseases and injuries, 1990-2015: A systematic analysis for the Global Burden of Disease Study 2015. Lancet. 2016;388(10053):1545-1602

[54] Makiala-MandandaS,Ahuka-MundekeS, Abbate JL, Pukuta-Simbu E, Nsio-Mbeta J, Berthet N, et al. Identification of dengue and chikungunya cases among suspected cases of yellow fever in the Democratic Republic of the Congo. Vector-Borne and Zoonotic Diseases. 2018;**18**(7):364-370

[55] WHO. A global brief on vector-borne diseases. Geneva, Switzerland, World Health Organization; 2014 Contract no: WHO/DCO/WHD/2014.1

[56] Curren EJ, Lindsey NP, Fischer M, Hills SL, St. Louis Encephalitis. Virus disease in the United States, 2003-2017. American Journal of Tropical Medicine Hygiene. 2018;**99**(4):1074-1079

[57] Center of Disease Control (CDC). 2022c. St. Louis encephalitis virus. Available at: https://www.cdc.gov/sle/faq/index.html. [Accessed on April 25, 2022]

[58] Thiberville SD, Nanikaly M, Laurence DM, Antoine N, Ernest AG, Pierre R, et al. Chikungunya fever: Epidemiology, clinical syndrome, pathogenesis and therapy. Antiviral Research. 2013;**99**:345-370

[59] WHO. 2020b. Vector-borne diseases. Available at: https://www.who.int/news-room/factsheets/detail/vector-borne-diseases. [Accessed: on April 20, 2022]

[60] O'Meara GF, Evans LF Jr, Gettnan AD, Cuda JP. Spread of *Aedes albopictus* and decline of *Ae. Aegypti* (Diptera: Culicidae) in Florida. Journal of Medical Entomology. 1995;**32**:544-562

[61] Romi R. History and updating of the spread of *Aedes albopictus* in Italy. Parassitologia. 1995;**37**:99-103

[62] Sikka V, Chattu VK, Popli RK, Galwankar SC, Kelkar D, Sawicki SG, et al. The emergence of Zika virus as a global health security threat: A review and a consensus statement of the INDUSEM joint working group (JWG). Journal of Global Infectious Diseases. 2016;**8**(1):3-15

[63] Mehrjardi MZ. Is zika virus an emerging TORCH agent? An invited commentary. Virology. 2017. Available at: 1178122X17708993. DOI: 10.1177/1178122X17708993. [Assessed on April 21, 2022]

[64] Boadle A, Brown T, Orr B. U.S., Brazil researchers join forces to battle Zika virus. Reuters. Brasilia. 2016. [Retrieved: March 4, 2016]

[65] Mahyoub JA, Rehman H, Saggu S, Murugan K, Panneerselvam C, Alrefaei MSS, et al. Insecticide susceptibility in larval populations of the West Nile vector Culex pipiens L. (Diptera: Culicidae) in Saudi Arabia. Asian Pacific Journal of Tropical Biomedicine. 2016;**6**(5):390-395

[66] Center of Disease Control (CDC). 2022d. West Nile dead birds. Available at: https://www.cdc.gov/westnile/dead-birds/index.html. [Accessed on April 25, 2022]

[67] Purse BV, Mellor PS, Rogers DJ, Samuel AR, Mertens PP, Baylis M. Climate change and the recent emergence of bluetongue in Europe. Nature Reviews Microbiology. 2005;**3**(2):171-181

[68] Howell PG, Verwoerd DW. Bluetongue virus. Virology Monographs. 1971;**9**:35-74

[69] Mann ME, Zhang Z, Rutherford S, Bradley RS, Hughes MK, Shindell D, et al. Global signatures and dynamical origins of the little ice age and medieval climate anomaly. Science. 2009;**326**(5957):1256-1260

[70] Campbell GL, Hills SL, Fischer M, Jacobson JA, Hoke CH, Hombach JM, et al. Estimated global incidence of Japanese encephalitis: A systematic review. Bulletin of the World Health Organization. 2011;**89**(10):766-774

[71] Hossain MJ, Gurley ES, Montgomery S, Petersen L, Fischer JSM, Panella A, et al. Hospital-based surveillance for Japanese encephalitis at four sites in Bangladesh, 2003-2005. American Journal of Tropical Medicine Hygiene. 2010;**82**(2):344-349

[72] Byam W, Lloyd LI. Trench fever: Its epidemiology and endemiology. Proceedings of Royal Society of Medicine. 1920;**13**:1-27

[73] Dworkin MS, Schwan TG, Anderson DE Jr, Borchardt SM. Tick-borne relapsing fever. Infectious Diseases Clinics of North America. 2008;**22**:449-468

[74] Christensen AM, Pietralczyk E, Lopez JE, Brooks C, Schriefer ME, Wozniak E. Diagnosis and management of *Borrelia turicatae* infection in febrile soldier, Texas, USA. Emerging Infectious Diseases. 2017;**23**:883-884

[75] Cutler S. Possibilities for relapsing fever reemergence. Emerging Infectious Diseases. 2006;**12**(3):369-374

[76] Rawlings JA. An overview of tick-borne relapsing fever with emphasis on outbreaks in Texas. Tex Medicine. 1995;**91**:56-59

[77] McCormick M. Rats, communications, and plague: Toward an ecological history. Journal of Interdisciplinary History. 2003;**34**(1):1-25

[78] Brites-Neto J, Duarte KMR, Martins TF. Tick borne infections in human and animal population worldwide. Veterinary World. 2015;**8**(3):301-315

[79] Alkishe AA, Peterson AT, Samy AM. Climate change influences on the potential geographic distribution of the disease vector tick *Ixodes ricinus*. PLoS One. 2017;**12**(12):e0189092

[80] Leighton PA, Koffi JK, Pelcat Y, Lindsay LR, Ogden NHJ. Predicting the speed of tick invasion: An empirical model of range expansion for the Lyme disease vector *Ixodes scapularis* in Canada. Journal of Applied Ecology. 2012;**49**:457-464

[81] Roy-Dufresne E, Logan T, Simon JA, Chmura GL, Millien V. Poleward expansion of the white-footed mouse (Peromyscus leucopus) under climate change: Implications for the spread of lyme disease. PLoS One. 2013;**8**:e80724

[82] Ellis J, Oyston PCF, Green M, Titball RW. Tularemia. Clinical Microbiology Reviews. 2002;**15**:631-646

[83] Morner T, Addison E. Tularemia, Ecology and animal health. In: Williams ES, Barker IK, editors.

Infectious Diseases of Wild Mammals. 3rd ed. Iowa: Iowa State University Press; 2001. pp. 303-313

[84] Rocklöv J, Dubrow R. Climate change: An enduring challenge for vector-borne disease prevention and control. Nature Immunology. 2020;**21**:479-483

[85] Paz S. Climate change impacts on West Nile virus transmission in a global context. Philosophical Transactions of the Royal Society B. 2015;**370**(1665):20130561

[86] Morin CW, Comrie AC, Ernst K. Climate and dengue transmission: Evidence and Implications. Environmental Health Perspective. 2013;**121**:1264-1272

[87] Smith K, Woodward A, Campbell-Lendrum D, Chadee D, Honda Y, Liu Q, et al. Human health: Impacts, adaptation, and co-benefits. In Climate Change 2014: Impacts, adaptation, and vulnerability. Part A: Global and sectoral aspects. Contribution of Working Group II to the Fifth Assessment Report of the Intergovernmental Panel on Climate Change. Cambridge University Press. 2014. pp. 709-754

[88] Mills JN, Gage KL, Khan AS. Potential influence of climate change on vector-borne and zoonotic diseases: A review and proposed research plan. Environment and Health Perspective. 2010;**118**:1507-1514

Chapter 8

Overview of the Main Species of Ticks and Animal and Human Tick-Related Diseases in the Caribbean, Particularly in Haiti

Max Francois Millien, Daphenide Saint-Louis and Daphnée Michel

Abstract

The Caribbean region faces a wide diversity of ticks and tick-borne diseases (TBDs) in animals and humans. But to date, these have been the subject of few studies, resulting in a relative lack of knowledge of their epidemiology, pathogenicity, and the best prevention and control methods. Ticks are hematophagous mites, which feed on the blood of mammals, birds, and reptiles. They are subdivided into two large families: the Ixodidae or hard ticks and the Argasidae or soft ticks. Each collection of blood by ticks from infected hosts can lead to their infection, which will contaminate other previously unharmed animals and contribute to the spread of tick-borne diseases caused mainly by bacteria, viruses, and parasites. It seems important to us to draw up a state of knowledge on ticks. Some long-known tick species like Rhipicephalus, Dermacentor, and Amblyomma and diseases like Anaplasmosis and Babesiosis deserve to be better studied, and others are yet to be identified for further research. The study consists of a review of the various documents published on this theme by Haitian and foreign researchers. The data are analyzed to assess the spatiotemporal distribution of ticks and identify the pathogenic germs they harbor and the various pathologies they induce in the Caribbean and Haiti.

Keywords: ticks, tick-borne diseases, epidemiology, research, Caribbean, Haiti

1. Introduction

Over the past five decades, infestations of human and animal populations by ticks have had a clear upward trend resulting in an increase in the prevalence of tick-borne diseases worldwide and particularly in the Caribbean. Several reasons have been put forward to explain these two facts. These are mainly climate change, new modes of land use linked to deforestation, the strong growth in demography in certain continents such as Asia, Africa, and America, the unprecedented development of the various livestock domestic animals, and their closer proximity to wild fauna in the case of extensive farming [1]. Indeed, ticks and wildlife are major reservoirs of pathogens that cause a whole range of infectious diseases transmitted by ticks, which can be of bacterial, viral, and parasitic origin such as anaplasmosis, babesiosis, borreliosis, ehrlichiosis, and rickettsioses in humans and animals [2].

These tick-borne diseases are distributed throughout the world, but fairly affected areas are found in subtropical and tropical regions such as the Caribbean [3]. Despite the existence of certain studies on this theme in the Caribbean, multidisciplinary teams must develop much more research activities on the epidemiology, ecology, and diagnosis of tick-borne diseases (TBDs) in animals and humans according to the "One Health" approach.

One of the advantages of this approach is that it helps to better understand the epidemiology of ticks and the diseases they transmit to animals and humans, given the fact that wild and domestic animals infested with ticks play an extremely important role in the epidemiology of zoonotic tick-borne diseases [2, 3]. In this study, we propose to present the epidemiological situation of these diseases in the Caribbean to take stock of the progress made and to identify new research projects to be initiated on this theme.

2. Classification and biology of ticks

Ticks are ectoparasites with an exoskeleton and articulated appendages that make them belong to the arthropod branch within the animal kingdom [4]. They belong to the order of mites and arachnids and are of major economic and health importance because of the direct and indirect losses they cause in animal production and their consequences on animal and human health. They are hematophagous mites that are found in almost all vertebrates throughout the world (mammals, birds, reptiles, etc.) but that only occasionally bite humans. During blood meals, they are capable of becoming infected if the hosts are already infected or of transmitting to their host pathogenic germs which induce animal (domestic livestock and wildlife) and human pathologies, some more severe than the others, and even become infected at the same time as they transmit pathogenic infectious agents [2]. They are known to be responsible for various disorders such as paralysis, allergy, abscess, anemia, immunosuppression, and deterioration of the skin at the bite site [5].

2.1 Classification

Ticks or Ixodida are divided into three families: the Nuttalliellidae having a single genus and a single species; the Ixodidae or hard ticks include 14 genera and 700 species; and the Argasidae or soft ticks include 5 genera and around 200 species. In the order of Ixodida, there are two superfamilies:

- Argasoidae (suborder Argasina) are ticks without a dorsal crest and are called "soft ticks."

DOI: http://dx.doi.org/10.5772/intechopen.106080

- This superfamily includes only one family which is, Argasidae, itself divided into two subfamilies: Argasinae, with a single genus: Argas, and Ornothidorinae which also has only one genus: Ornothodoros;

- the Ixodoidae (of the suborder Ixodina) known as "hard ticks" which include 14 genera and about 700 species), and the Argasidae or soft ticks (which include 5 genera and about 200 species). It has been estimated that nearly 10% of the 900 known tick species can transmit pathogens [6].

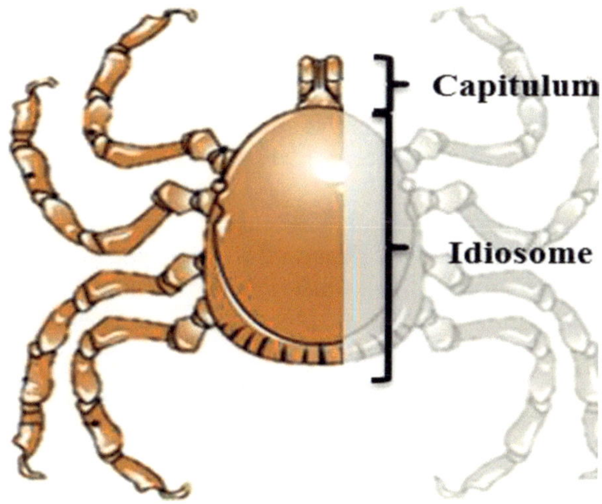

Figure 1.
Body division of an Ixodidae. Source: [3].

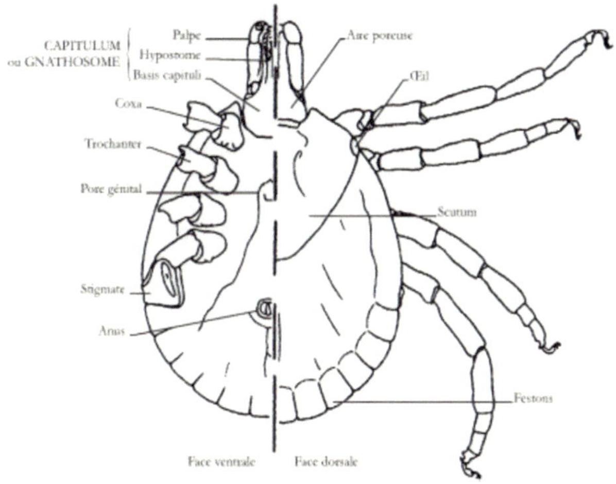

Figure 2.
General morphology of a tick. Source: [9].

Ticks are known to be among the arthropods, the vectors that transmit the greatest variety of pathogenic microorganisms, including bacteria, viruses, protozoa, and helminths [7].

The Ixodoidae superfamily is the best known and the most studied, because it includes most of the ticks that are adapted to domestic animals and/or that can bite humans. They infest different species of domestic animals and wild fauna in which they cause significant economic losses to breeders around the world and especially in the Caribbean. On the other hand, some species of these ticks are capable of accidentally infesting humans and causing severe infectious diseases such as Lyme disease due to the bacterium *Borrelia burgdorferi*, thus disrupting the functioning of health systems, public, and social security in countries other than those of the Caribbean.

The superfamily (Ixodoidae) contains two families: Ixodidae and Amblyommidae. The Ixodidae family has only one subfamily (Ixodinae) and one genus: Ixodes. As for the second family, that of the Amblyommidae, it includes several genera: Haemaphysalis, Amblyomma, Dermacentor, Hyalomma, and Rhipicephalus Boophilus.

The Ixodidae or hard ticks are so called because of their structure and conformation because they have a hard plate or crest on the dorsal side of their body, while the Argasidae or soft ticks have a soft integument devoid of a dorsal crest.

Ixodidae can survive much more easily than Argasidae in unfavorable conditions to be able to go long without feeding and to adapt to several hosts. Hard ticks are also more prolific, because they lay more eggs and are less vulnerable to having fewer natural enemies than soft ticks [8].

2.2 Morphology of ticks

Ticks are mites whose body is divided into two parts, the capitulum or gnatostome, which mainly bears the mouthparts and the idiosome on which the legs are attached. Unlike insects which have three pairs of legs, they have four. A global description of an Ixodidae tick looks like as shown in **Figures 1** and 2.

The morphology of ticks is directly related to their hematophagous lifestyle. The rostrum, carried by the capitulum, is similar to a blood collection instrument composed of two chelicerae and a hypostome which will penetrate the tissues of the vertebrate host for their blood meal. The idiosome is covered almost entirely by an extensible cuticle which allows it to expand and store blood. The males of adult hard ticks absorb little or no blood and, on the contrary, do not have an extensible cuticle but a scutum that covers the entire dorsal surface of its idiosome, preventing the distention of the idiosome and thus not allowing only slight dorsoventral dilation. Also, they absorb little or no blood. On the contrary, in female Ixodidae, the scutum only covers a small part of the body on the dorsal side and is located at the base of the capitulum [4].

2.3 Life cycle and main hosts

Ticks are large mites that present three stasis separated by real metamorphoses: larva, nymph, and male or female adult, corresponding to the mature stasis. The larva is easily distinguished by its small size and the number of its pairs of legs because, at this stasis, it has only three pairs. The nymph is distinguished from the female by the absence of a genital pore and porous areas in species that have them **Figure 3**.

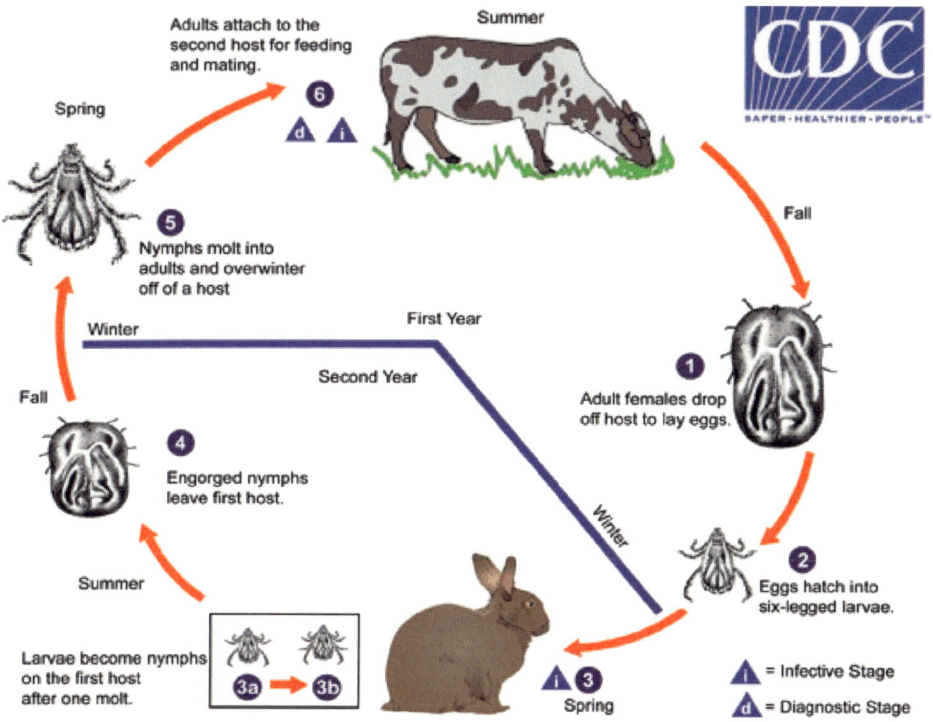

Figure 3.
Life cycle of the two-host ixodid (hard) ticks [10].

2.3.1 Reproduction of ticks

Ixodidae mating usually occurs on the host. At the end of their meal, the fertilized female detaches herself and falls to the ground where she lays her eggs. Hard ticks are reputed to be most often exophilic, that is to say, they live in open biotypes such as forests, pastures, savannahs, or grasslands. But some species are endophilic because they are found in protected habitats such as burrows or nests.

After hatching from the eggs, the larvae quickly seek out a host to feed on blood. Then, they detach and fall to the ground to undergo a metamorphosis into nymphs that can last from 2 to 8 weeks depending on the species and the climatic conditions. As for the phase of transformation into adults, it is generally longer because it extends up to 20 to 25 weeks. Environmental and climatic conditions influence the length of the life cycle of Ixodidae. The males feed little; only females take a large enough blood meal to ensure spawning [4].

The Ixodidae are by far the most important family of ticks in human and veterinary medicine because they include 80% of species in the world and have a triphasic life cycle: larva, nymph, and adult. Each stage of the vector tick cycle corresponds to a different vertebrate host, bites it, attaches to it, and takes a unique blood meal.

As for soft ticks (Argasidae), they have several nymphal stages before metamorphosis into adults. They generally live in dry areas and have great resistance to desiccation and fasting. Most species of Argasidae are endophilic; the distribution of species as well as the diseases transmitted is often limited. Generally, the parasitic stages feed on the host several times over a short period (from a few minutes to a few hours), ingest a relatively small amount of blood per meal, and then return to their nest [7].

In some species, the sting is painful unlike that of the Ixodidae which is often painless. The main soft ticks are of the genus Ornithodoros, vectors of the agents of relapsing fever and African swine fever [2].

2.3.2 Finding the host

In the process of finding the host and a sexual partner, the sensory organs of ticks play a vital role. These are also used in the assessment of climatic conditions. In addition to the bristles distributed over the entire body of ticks endowed with mechano-proprioceptive or chemoreceptor functions and the eyes in certain species of ticks, ticks have a very particular organ, the Haller organ, which is sensitive, among other things, to the degree hygrometry and pheromones, allowing ticks to locate their host by detecting the CO_2 that this emits as well as the heat and metabolites that it releases [11].

Exophilic ticks search for their host in two ways. They can practice passive waiting by climbing on the vegetation at a variable height according to the species or the stages and wait for the passage of a host with their forelegs raised to be able to cling to it. On the contrary, they can resort to an attack strategy by leaving their habitat and going toward hosts that are within their reach to attract them by different stimuli that they emit. Some tick species can use both strategies. The choice of one or the other can then vary according to the stage within the same species.

Ticks have preferential attachment sites on vertebrate animals, which vary according to species and sometimes according to stages within the same species. They can bite humans all over the body but are often found on the head, neck, or groin. But it is important to know that there is no tick specific to humans who always become infected only accidentally when they share the biotope of tick-infested animals [2].

2.3.3 Methods of transmission of infection by ticks

A vector can be defined as a hematophagous arthropod responsible for the active biological transmission of an infectious agent which can be viral, bacterial, or parasitic. The vectorial transmission of this infectious agent implies that it is taken by the vector during its blood meal and that it retransmits it in favor of another blood meal taken from a new host. For there to be the transmission, the infectious agent must remain alive in the vector between the two blood meals, either by transstadial transmission or by transovarial transmission. The biological transmission of an infectious agent implies the existence within the vector of a phase of its life cycle (multiplication or antigenic modifications).

Biological transmission of an infectious agent differs from simple passive transport, which refers to mechanical transmission. The tick, as a vector, has an active role in the contact between the infectious agent and the vertebrate host. A distinction is made between the main vectors of infectious agents which can be responsible on their own for an endemic disease, from those known as secondary or even accidental. These different vectors may very well vary from one region to another for the same infectious agent [2].

Hard ticks typically have a type of transstadial transmission that occurs when a vector retains a pathogenic infectious agent in its body as it transitions from one developmental stage to another, i.e. when the vector tick is infected at the larval stage and that this infection can persist when the larva transforms into a pupa and then into an adult form. This type of transmission is the necessary condition for a tick to be a vector of an infectious agent. Although the nymphs and adult larvae represent the main vectors of pathogenic germs, the larvae also play a role in the transmission of infectious agents by transmission [9].

In the light of the phenomena of transstadial transmission and or transovarial transmission, we can explain the problem of co-infections by various microorganisms which are often observed in ticks, making the diagnosis and treatment of certain induced pathologies more difficult [12]. However, reports of co-infections in humans in the Caribbean are rare.

2.3.4 Impact of climate change on ticks

Many facts establish a correlation between the occurrence of tick-borne diseases and climate change. It is indeed known that climatic factors influence the multiplication of ticks, their life cycle, the seasonal variation of their activity and behavior as well as population dynamics. According to data published by the National Climate Assessment of the United States, there would be a 2°C increase in the average annual temperature during the middle of the twenty-first century (2036–2065), which could lead to a 20% increase in the number of cases of Lyme disease in the United States in the decades to come.

The parameter of geography has to be taken into consideration in the study of the distribution of ticks. Indeed, each species tends to present a particular geographical distribution, and the diseases transmitted by ticks which are both vectors and reservoirs of germs are considered geographical diseases. But still, with climate change, some species of ticks are starting to become more and more cosmopolitan.

3. Physical framework for the study of ticks

3.1 Main physical, geographical, and climatological characteristics of the Caribbean

The Caribbean is a geographical unit formed by the Antilles and part of the circumference of the Antilles Sea. It is also called the Caribbean or Caribbean space and represents a region of the American continent that includes the Caribbean Sea, islands, some of which are surrounded by the Caribbean Sea, and other islands which border both this sea and the North Atlantic Ocean and surrounding coasts. It is located, for the most part, on the Caribbean Plate, which is a region with more than 700 islands, islets, and reefs. Generally, when we talk about the Caribbean Community, the first agglomerations that come to mind are the Greater Antilles and some Lesser Antilles.

The Greater Antilles include Cuba, Dominican Republic, Haiti, Jamaica, and Puerto Rico. The Lesser Antilles form part of the Antilles Archipelago which extends from Puerto Rico to the coast of Venezuela. The most representative islands of this group are the Virgin Islands, Saint-Martin Island, Anguilla, ABC Islands: Aruba, Bonaire and Curaçao, Antigua and Barbuda, Saint Kitts and Nevis, Guadeloupe, Dominica, Saint Vincent, and the Grenadines, Grenada, Martinique, Saint Lucia, and Barbados.

This region is often confronted with cyclones and is often the scene of earthquakes, particularly in Haiti. Some areas of this region are also prone to volcanism.

Its temperature levels are typical of a tropical climate and are softened by sea breezes and by altitude. There are two distinct seasons: the dry season which is cool and which extends from December to June, and the wet season which is hot from June to December, known as the hurricane period. It is generally characterized by stormy rains and the passage during the summer of cyclones with destructive effects on agriculture, livestock, various infrastructures (irrigation canals, bridges, etc.), and human life.

3.2 Some aspects of health situation in the Caribbean region

In the Caribbean region, because of the glaring differences between the animal health and public health systems of certain countries of this region, there are great differences between them in terms of overall health and in their capacity for epidemiological surveillance of ticks and tick-borne diseases.

All the islands of the Greater Antilles as well as a good number of the Lesser Antilles are members of the Caribbean Animal Health Network (CaribVET) or are developing cooperative relations with this network in terms of animal health. The Caribbean countries led by CIRAD established this network some 20 years ago with the technical support of certain regional organizations such as IICA and APHIS/USDA. The aim was to network the animal health and veterinary public health services of the countries of the Caribbean region including, among others, Haiti, the French, American, British and Dutch West Indies, as well as Cuba and the Dominican Republic.

CaribVET is today the only animal health and veterinary public health network in the Caribbean, recognized by regional (CARICOM, CAHFSA, and IICA) and international (OIE, FAO, PAHO, and USDA) bodies which are themselves involved in its governance. It also brings together the veterinary services of 34 countries/territories, 7 universities and research and development centers, totaling 48 members.

The purpose of this network is to provide an operational response to health issues and the emergence of animal and zoonotic diseases in this complex region made up of more than 30 territories of quite varied size, level of development, and political status.

4. Ticks and transmission of diseases to animals and man

4.1 Pathogenic power of ticks

Ticks have a pathogenic role that is both direct and indirect. The direct role results in blood loss, paralysis, toxicosis, tropical dyshidrosis, allergic reactions, and wounds which can then become superinfected [4]. The indirect role is exerted through the transmission of infectious agents to their vertebrate hosts. They are vectors that transmit the greatest diversity of infectious agents in the world and are the second vector after mosquitoes, as far as human public health is concerned [13]. They are involved in the transmission of pathogens to both humans and animals [3].

4.1.1 Vectorial capacity of ticks

The vector competence of an arthropod, i.e. its ability to transmit an infectious agent, is an essential condition for it to be considered as a vector of this agent, without however being sufficient. Indeed, other parameters such as the abundance of the vector, the dispersal capacity of the vector via its hosts, the ecological preferences of the vector in terms of habitat, host, and activity, the trophic and ecological preferences of ticks, the age or the stasis of the vector are also important to make an arthropod the vector of an infectious agent or to determine its vectorial capacity [14].

Ticks are considered to be good vectors of infectious agents whose contact with their hosts can be either prolonged in hard ticks which have long and copious blood meals or repetitive in soft ticks with meals of very low volume [4]. In both cases, the exchange of infectious agents is facilitated. Ticks can therefore have broad to very broad host spectra, thus promoting the circulation of infectious agents [15]. In addition, during their metamorphoses, they undergo few rearrangements thus favoring, most often, the preservation of infectious agents between each stasis (transstadial transmission) [15].

It is recognized that the most important ticks in terms of their ability to transmit diseases to ruminants in the Caribbean islands are as follows: Amblyomma variegatum, a vector of heartwater and associated with acute dermatophilosis; Amblyomma cajennense, a potential vector of heartwater; Boophilus microplus, vector of babesiosis and anaplasmosis [16].

4.1.2 Tick infection and co-infection process

As the biological cycles of ticks can be very long, they also serve as reservoirs of infectious agents in nature [4]. As they can be transported by their hosts, especially birds, over very long distances, they cause a very significant dispersion of the infectious agents they carry [17, 18]. The fact of having several meals during their life allows them to undergo co-infections, which are to say to harbor at the same time several infectious agents that can be transmitted to susceptible hosts.

The probability of having an infecting tick vector is all the greater when the concentration of infectious agents is high and their presence in the blood of the host is prolonged. However, it has been observed that infectious agents present in an infected tick can pass to a healthy tick during the co-meal phenomenon in the absence of viremia or bacteremia in the host, even in the presence of targeted antibodies of the host [4]. This is because the bite site undergoes, under such conditions, significant local pharmacological modifications linked to the injection of active substances contained in the saliva of ticks promoting the action of certain infectious agents to contaminate naive ticks [19]. However, co-meal transmission is only possible if the infected tick and the naive tick are attached a short distance from each other.

Such behavior by facilitating both blood feeding and mating contributes to the transmission of pathogens from an infected tick to an uninfected one sharing the same blood meal. Ticks can then become infected with pathogenic germs during co-feeding without the host being bacteremia or viremic, i.e. carrying these germs [2, 20]. Tick co-infections are well known in the Caribbean region, especially in dogs. This fact has been observed in several countries in the region such as Grenada, St Kitts, and Haiti. By way of illustration, let us present some results of co-infections in Haiti and Cuba.

In Haiti, co-infection in dogs with two or more PTBs was detected using serology (20.0%) and molecular methods (10.6%). The most common co-infection involved Dirofilaria immitis and B. Vogeli (3.4%), followed by *D. immitis* and *E. canis (1.9%), D. immitis* and *H. canis (1.4%), E. canis* and *H. canis (1.4%), H. canis* and *Acanthocheilonema recondite (1.0%), D. immitis* and *Anaplasma platys (0.5%), E. canis* and *B. vogeli (0.5%),* and *H. canis* and *A. platys* (0.5%).[21].

In Cuba, different co-infections have been identified with:

- three haemoparasites (B. bigemina, A. marginale, and B. bovis) in 12.0% of water buffaloes, co-infections with B. bovis and A. marginale being the most common (26.0%). %) followed by B. bovis/B. bigemina (20.0%) and A. marginale/B. bigemina (24.0%), suggesting the potential positive interaction between these pathogens.

- B. caballi and T. equi in 20.0% of horses tested.
- A. ovis and B. ovis and between A. ovis and E. ovis
- A. ovis and B. motasi in sheep by microscopic examination of blood smears [22].

4.1.3 Other ways of transmission of a tick-borne disease

Not all tick-borne diseases are transmitted exclusively by tick bites. This is particularly the case for certain pathogenic germs such as Bartonella, Francisella, and Coxiella which use other routes for the transmission of infectious agents. Indeed, the transmission of Q fever or coxiellosis does not generally occur by infected tick bite but by inhalation of contaminated dust (feces of infected ticks) and by contact with infected secretions, including milk, and also the placenta, aborted small ruminants. It has been shown that in the case of certain human epidemics cataloged as tick-borne

Pathogens	**Caribbean distribution**	**Infection (TBDs)**	**Vector***	**Reported host**	**Detection method**
Anaplasma marginale	Antigua, Barbados, Cuba, Dominica, Dominican Republic, Grenada, Guadeloupe, Haiti, Jamaica, Martinique, Montserrat, Puerto Rico, St Kitts and Nevis, St Lucia, St Martin, St Vincent, Trinidad	Anaplasmosis	*Rh. microplus*	Cattle	Serology
Anaplasma phagocytophilum	Puerto Rico	Anaplasmosis (Granulocytic Anaplasmosis)	*Ixodes* spp	Dogs	Serology
Anaplasma platys	Cuba, Grenada, Haiti, St Kitts, Trinidad,	Anaplasmosis (Canine Cyclic Thrombocytopenia)	*Rh. sanguineus, A.cajennense*	Dogs	Serology/ Molecular biology
Bartonella vinsonii subspp. *Berkhoffii*	Grenada, Martinique	Canine and human endocarditis	*Rh.sanguineus*	Dogs	Serology/ Molecular biology
Borrelia burgdorferi sensu lato	Cuba	Borreliosis (Lyme disease)	*A.cajennense*	Humans	Serology
Borrelia Relapsing Fever group (Borrelia hermsii)	US Virgin Islands	Borreliosis (Relapsing fever)	*Ornithodoros* spp	Humans	Serology

Pathogens	Caribbean distribution	Infection (TBDs)	Vector*	Reported host	Detection method
Coxiella burnetii	Puerto Rico	Q fever	*D. nitens*	Cattle	ND
Ehrlichia canis	Aruba, British West Indies, Grenada, Haiti, Puerto Rico, St Kitts and Nevis, Trinidad, Turks and Caicos Islands	Ehrlichiosis (Canine Monocytic Ehrlichiosis)	*Rh. Sanguineus*	Canids, Cats	Serology/ Molecular biology
E. canis or closely related species	Dominica, Grenada, Montserrat, St Kitts and Nevis	Ehrlichiosis	ND	Cattle, Sheep, Goats	Molecular biology
Ehrlichia ruminantium	Antigua, Guadeloupe, Marie Gualante	Ehrlichiosis (Heartwater, Cowdriosis)	*A. variegatum*	Cattle	Serology
Panola Mountain Ehrlichia sp.	Dominica, St Kitts	Ehrlichiosis	*A. variegatum*	Domestic animals	Molecular biology
Candidatus Mycoplasma haematoparvum	Trinidad	Hemotropic mycoplasmosis	*Rh.sanguineus*	Dogs	Molecular biology
Mycoplasma haemocanis	Trinidad	Hemotropic mycoplasmosis	*Rh. sanguineus*	Dogs	Molecular biology
Mycoplasma haemofelis	Trinidad	Hemotropic mycoplasmosis	*Rh. sanguineus*	Cats	Molecular biology
Candidatus Mycoplasma haemominutum	Trinidad	Hemotropic mycoplasmosis	*Rh. sanguineus*	Cats	Molecular biology
Mycoplasma wenyonii	Cuba	Hemotropic mycoplasmosis	ND	Cattle	ND
Mycoplasma ovis	Cuba	Hemotropic mycoplasmosis	ND	Sheep	Serology
Rickettsia africae	Antigua, Dominica, Guadeloupe, Martinique, Montserrat, St Kitts and Nevis, St Lucia, U.S. Virgin Islands	Rickettsiosis (African tick bite fever)	*A.variegatum*	Humans, Cattle, Goats, Sheep	Serology/ Molecular biology
Rickettsia conorii	Guadeloupe	Rickettsiosis (Mediterranean spotted fever)	*A. variegatum*	Humans	Serology
Rickettsia felis	Dominica, St Kitts	Rickettsiosis	*Ctenocephalides felis*	Cats	Molecular biology

Pathogens	Caribbean distribution	Infection (TBDs)	Vector*	Reported host	Detection method
Rickettsia typhi	Puerto Rico	Rickettsiosis (Murine typhus)	*Xenopsylla cheopis*	Rodents	ND
Babesia bigemina	Antigua, Barbados, Cuba, Dominica, Dominican Republic, Grenada, Guadeloupe, Haiti, Jamaica, Martinique, Montserrat, Puerto Rico, St Kitts and Nevis, St Lucia, St Martin, St Vincent, Trinidad	Babesiosis	*Rh. microplus*	Cattle	Serology/ Molecular biology
Babesia bovis	Antigua, Barbados, Cuba, Dominica, Dominican Republic, Grenada, Guadeloupe, Haiti, Jamaica, Martinique, Montserrat, Puerto Rico, St Kitts and Nevis, St Lucia, St Martin, St Vincent, Trinidad	Babesiosis	*Rh. microplus*	Cattle	Serology/ Molecular biology
Babesia caballi	Grenada, Guadeloupe, Martinique, Montserrat, St Kitts and Nevis, Trinidad	Piroplasmosis	*D. nitens*	Equids, Goats, Sheep	Serology/ Molecular biology
Babesia (canis) rossi	Montserrat	Babesiosis	*Rh.sanguineus, Rh. turanicus*	Goats	Molecular biology
Babesia (canis) vogeli	Dominica, Grenada, Haiti, Montserrat, St Kitts and Nevis, Trinidad	Babesiosis	*Rh. sanguineus, Rh. turanicus*	Dogs, Cats, Cattle, Sheep, Goats	Serology/ Molecular biology

Pathogens	Caribbean distribution	Infection (TBDs)	Vector*	Reported host	Detection method
Babesia gibsoni	Dominica, St kitts	Babesiosis	*Rh.sanguineus, Rh.turanicus*	Dogs, Cattle, Sheep, Goat, Equids	Molecular biology
Babesia vulpes	Montserrat	Babesiosis	ND	Goats, Sheep	Molecular biology
Hepatozoon canis	Aruba, Grenada, Haiti, St Kitts, Trinidad	Hepatozoonosis	*Rh. sanguineus*	Dogs	Serology/ Molecular biology
Theileria equi	Dominica, St kitts and Nevis, Trinidad	Piroplasmosis	*D. nitens*	Equids, Cattle, Sheep, Goats	Serology/ Molecular biology
Theileria mutans	Cuba, Guadeloupe, Martinique	Theileriosis	*A. variegatum*	Cattle	Serology (IFA)
Theileria parva	Guadeloupe	Theileriosis	*A. variegatum*	Cattle	ND
Theileria sp. B15a	Grenada	Theileriosis	ND	Cattle	Molecular biology
Theileria sp. NG-2013a	Nevis	Theileriosis	ND	Goats	Molecular biology
Theileria sp. OT3	Montserrat	Theileriosis	ND	Sheep, Goats	Molecular biology
Theileria sp. YW-2014	St Kitts	Theileriosis	ND	Equids	Molecular biology
Theileri velifera	Guadeloupe	Theileriosis	*A. variegatum*	Cattle	Serology
African swine fever	Cuba, Dominican Republic, Haiti	African swine fever	*Ornithodoros* spp	Swine	ND
Estero Real	Cuba	ND	*C. tadaridae*	ND	Isolation
Hugues Virus	Cuba, Trinidad	ND	*C. denmarki*	Seabirds	Isolation
Soldado Virus	Trinidad	ND	*Carios* spp	Seabirds	Isolation
Wad Medani Virus	Jamaica	ND	*A.cajennense*	ND	Isolation

Source: [31].

Table 1.
Tick-borne pathogens and suspected tick vectors reported within the Caribbean.

diseases, such mites may not be directly involved as vectors. Examples include human piroplasmosis which can occur through blood transfusion [23].

- tick-borne encephalitis, the virus of which can be transmitted to humans by drinking infected milk;
- Crimean-Congo hemorrhagic fever, the virus of which can be contracted by handling the carcasses of infected animals.

4.2 Tick species identified

4.2.1 Origin of ticks

In the Caribbean, the tick fauna is made up, on the one hand, of endemic species and, on the other hand, of exotic species which have been introduced there by different routes: movement of animals in the region, movement of birds migratory or non-migratory from North, Central, or South America, occupation of this region by settlers who came with infested cattle and dogs from Europe, Africa, and Asia [24, 25]. A total of 56 species of ticks have been recorded in the Caribbean, belonging to 10 genera and 2 families (Argasidae and Ixodidae) including 15 species of Ornithodoros, 10 species of Antricola, 17 species of Amblyomma, 3 species of Argas, Ixodes, and Rhipicephalus, 2 species of Haemaphysalis, and 1 species each of Parantricola, Dermacentor (Anocentor), and Aponomma [26].

Because of their impact on health, the most studied tick species in the West Indies are those associated with the transmission of tick-borne diseases to livestock or pets. However, even if the tropism of ticks for certain hosts is well documented, it is observed that many of them can parasitize different species of hosts, including humans, opening the way to the occurrence of zoonotic diseases [27, 28]. Other species of ticks, present in wildlife, also exist in the Caribbean without arousing much concern and interest to date. As the majority of emerging diseases come from wildlife reservoirs, characterization of the diversity and ecology of ticks present in these wild environments should be addressed [29, 30]. The tick species described in this review are significantly implicated in the epidemiology of animal tick-borne diseases (**Table 1**).

4.2.2 Diseases transmitted by Ixodidae ticks to cattle

- *Rhipicephalus (Boophilus) microplus*

It is the tropical cattle tick considered to be the most important in the world. Although primarily associated with cattle, *Rh. microplus* can feed on a variety of hosts among domestic animals in the Caribbean such as horses, donkeys, goats, sheep, buffaloes, pigs, dogs, and also in some wild animals [32]. According to some cases of human infestation by *Rh. microplus* have sometimes been detected in humans [33]. But these *Rh. microplus* ticks are single host, i.e. they can stay on the same host throughout their life cycle. They are well distributed in the Caribbean, both in the Greater and the Lesser Antilles [16].

This cattle tick is mainly involved in the transmission of bacteria and protozoa such as *Anaplasma marginale, Anaplasma centrale, Babesia bigemina, Babesia bovis, and Theileria spp* [34].

In almost all Caribbean countries with a fairly large livestock population, significant economic losses of several million US dollars have been recorded, as was the case in Puerto Rico.

- *Amblyomma variegatum*

Unlike *Rhipicephalus microplus*, the tropical tick *A. variegatum* is a tick with several hosts which are mainly ruminants (cattle and goats). It causes deep skin lesions through their mouthparts which can facilitate the development of secondary infections leading to acute dermatophilosis. It transmits pathogenic bacteria such as *Ehrlichia spp. and Rickettsia spp*., and protozoa such as *Theileria spp*. [27, 35–37]. It also caused significant losses during the 1990s which amounted to several million US dollars in the English Lesser Antilles.

The tick was first introduced to Guadeloupe and for a time remained confined to Guadeloupe, Antigua, and Martinique until the 1960s. But by the 1980s, it had spread to 18 islands in the Caribbean in the late 1980s. So far, the distribution of *Amblyomma variegatum* in the Caribbean is restricted to the Lesser Antilles [16, 38].

- *Amblyomma cajennense*

A. cajennense, a tick with multiple hosts, has also been reported in several Caribbean countries such as Cuba, Jamaica, and Trinidad. However, due to deficiencies in the epidemiological vigilance system for ticks and tick-borne diseases in some Caribbean countries, it cannot be said with certainty that it is limited only to these three countries. It is suspected to be the vector of the agents *Ehrlichia spp., Rickettsia spp*., and equine piroplasms [39].

- *Dermacentor (Anocentor) nitens*

The tropical horse tick is a single-host tick that primarily parasitizes equines. *D. nitens* is suspected of being the vector of *Babesia caballi and Theileria equi,* the two causative agents of equine piroplasmosis [40].

- *Rhipicephalus annulatus*

The existence of *R. annulatus* has not been established with certainty in the Caribbean because according to some authors, there is a risk of confusion about this tick with Rh. microplus because of their morphological similarities.

- *Rhipicephalus sanguineus*

It is a brown tick commonly found in dogs, but it has three hosts and is considered widespread in the Caribbean and often parasitizes dogs [24–26]. It is prevalent in several Caribbean countries including Haiti and plays an important role in both animal health and public health due to its ability to transmit various Ehrlichia, Anaplasma, Rickettsia, and Babesia pathogens [28].

4.2.3 Tcks and tick-borne diseases in pets

The most common ticks and bacterial and protozoan diseases transmitted by ticks in dogs in the Caribbean are as follows:

- tropical canine pancytopenia (caused by *Ehrlichia canis*),
- canine cyclic thrombocytopenia (*Anaplasma platys*), and
- canine babesiosis (*Babesia canis*) [21, 37, 41].

- **Canine babesiosis** is one of the most commonly reported infections and is often caused by *Babesia canis vogeli and Babesia gibsoni*. It has been detected in Haiti, Grenada, and St Kitts.

Another hemiparasite, *Hepatozoon canis*, is acquired by ingesting infected ticks; it has been reported in certain Caribbean islands such as Cuba, Grenada, Saint-Kitts, and Trinidad [31, 42–44].

The above pathogens are usually transmitted or associated with the bite or ingestion of *R. sanguineus* ticks. It should be noted that natural infections with *Rickettsia amblyommatis* and *Rickettsia rickettsii* have been detected in *R. sanguineus* [45].

- *Babesia gibsoni* **infections** are usually acute and characterized by anorexia, fever, hepatomegaly, splenomegaly, and pallor. Some animals, after recovery, however, remain carriers. Other Babesia infections reported in mammalian hosts include *B. Gibson and Babesia vogeli* in cats. [3, 46].

4.2.4 Ticks as vectors of human pathogens of bacterial and protozoan origin in the Caribbean

R. rickettsii is a highly pathogenic agent causing Rocky Mountain spotted fever (RMSF) in humans. This has been detected in Belize and some Caribbean countries. Its symptoms are high fever, violent headaches, and exanthema. A less severe rickettsia, African tick-bite fever, caused by *R. africae*, has been reported from the eastern Caribbean [47]. This TBP was probably introduced into the Caribbean with cattle infected with *A. variegatum* imported from Senegal more than 200 years ago. Although a small number of clinical human cases are reported, R. africae and its human-infecting vector, *A. variegatum*, are widely distributed in the Caribbean [48].

4.2.5 Human and animal viruses transmitted by ticks in the Caribbean

Several tick-borne viruses, particularly the Argasidae, have been identified in ticks in the Caribbean region, including Hughes virus in the Ornithodoros capensis (Ornithodoros denmarki and Soldado virus in *Ornithodoros capensis* of Trinidad [49, 50], and *O. denmarki* in Cuba [51], Estero real virus in *Ornithodoros tadaridae* ticks in Cuba [52], and Wad Medani in A. cajennense from Jamaican ticks [53] .

The African swine fever virus, which raged in Haiti, the Dominican Republic, and Cuba in the 1980s, was detected in Ornithodoros ticks [54]. However, there was no evidence that these had participated in the spread of ASF under natural conditions because this virus was not identified in any of the 350 Ornithodoros puertoricensis ticks collected in the Dominican Republic and Haiti. Ticks were unable to acquire and transmit this virus transstadially and transovarially under laboratory conditions [55, 56].

4.3 Veterinary and health importance of ticks

4.3.1 Direct and indirect impact of ticks on human and animal health

The impact of ticks is quite significant in both animal and human health. Their natural hosts are normally wild and domestic animals. But, when man finds himself in the biotopes of ticks, he can become an accidental host, [57]. The medical importance of ticks must be understood on the one hand, through their direct role on the health of their host, which is dependent on their behavior as hematophagous ectoparasites, and, on the other hand, through their indirect action as carriers of a large number of direct pathogens.

4.3.2 Direct health impacts of ticks

The direct effects of ticks on the health of the host are particularly observed in animals where they cause significant economic losses for the livestock industries. Some species of ticks such as those of the genus Amblyomma have the ability to take about 4 ml of blood during their blood meal, thus creating cases of blood loss, anemia, and significant weakening of the animal infested. When the tick is attached to a given host, the rostrum will penetrate the skin of the host, which can often transform this bite into a cutaneous wound capable of causing superinfections, the introduction of opportunistic pathogens or the formation of abscesses. A bacterium like *Dermatophilus congolensis* responsible for bovine dermatophilosis is widely associated with cases of infestation of the animal by *A. variegatum* ticks, whose bite sites are entry routes for the pathogen. The leather industries are particularly affected by the poor quality of animal skin linked to tick infestation. Similarly, domestic animal infestations can result in a significant drop in productivity for livestock industries for meat and milk [7]. Another consideration is that tick saliva is made up of several different molecules in the feeding site during the blood meal with possible consequences of reduced host immune system, transmission of pathogens, and even allergies due to the presence of certain molecules which can prove to be toxic for the host (animal or human) leading to cases of toxicosis or serious paralysis [5, 58].

4.3.3 Indirect health impacts of ticks

Ticks are the most important vectors in animal health, and they are placed before the mosquito. In human health, they come second to mosquitoes. They ensure active transmission (mechanical or biological) of an infectious agent from one vertebrate to another vertebrate. Globally, ticks are responsible for transmitting the widest range of pathogens. They transmit microorganisms responsible for bacterial (Lyme borreliosis and rickettsioses) or parasitic (babesiosis and theileriosis), or even viral (tick-borne encephalitis) diseases. The microorganisms responsible for these diseases constitute a major risk for both human and animal health. Ticks are capable of parasitizing many host species, affecting wildlife, livestock, pets, and even humans.

Without having precise figures, it is estimated that the financial costs related to ticks and tick-borne diseases are considerable in public and animal health in the Caribbean [7].

5. New perspectives for the development of a better system for the epidemiological surveillance of ticks and tick-borne diseases

As has been demonstrated, the Caribbean is a region at risk for the proliferation of vector-borne diseases. Such a situation is explained by favorable conditions for vectors, increasingly important intercontinental exchanges, animal movements, and a strong presence of migratory birds in this region. Also, it is important to develop increasingly efficient epidemiological surveillance systems for ticks and tick-borne diseases. But conventional diagnostic techniques have proven ineffective for a true assessment of the extent of these pests and diseases. New diagnostic tools are needed to draw up a more in-depth inventory of pathogens of medical and veterinary interest present in ticks in the Caribbean region. The use in Guadeloupe and Martinique of high-throughput RNA sequencing techniques extracted from ticks collected in the Caribbean gives great hope for an exhaustive inventory of pathogens (bacteria, parasites, and viruses) present in the region.

The analyses thus carried out revealed a great diversity of pathogenic germs within the samples collected which would not be detected by conventional detection techniques. For example, they revealed:

- The presence of four viruses belonging to new viral genera recently described and associated with arthropods.

- The constitution of a list or directory of pathogens transmitted by ticks requiring health surveillance in the Caribbean and which can make it possible to develop a high-throughput screening system for infectious agents applicable to the entire Caribbean area.

- Important data on the epidemiological situation of the Lesser Antilles with regard to pathogens transmitted by ticks, i.e. 45 bacteria, 17 parasites, and 31 viruses potentially from the sampling of the study carried out in Guadeloupe and Martinique.

Despite the wide variety of ticks and pathogens transmitted, few studies have been devoted to assessing the diversity of tick species and the incidence of tick-borne diseases in humans [59, 60]; it is now necessary to actively monitor the spread of important pathogens transmitted by ticks other than those which have traditionally been transmitted such as *A. marginale, Ehrlichia ruminantium, Babesia. bovis, Babesia bigemina, Babesia caballi, and Theileria equi*, due to their socio-economic impact [61, 62].

The surveillance and control strategy against ticks and tick-borne diseases in the Caribbean benefit from adopting the "One Health" approach, which implies greater collaboration between veterinary health and environmental professionals for a comprehensive approach of health because over the past few decades, changes in ecosystems have contributed to making wildlife the main reservoir of pathogenic germs. The "One Health" approach should enable the improvement of the diagnostic process, the acceleration of treatment decisions, and the adoption of appropriate prevention and control protocols.

6. Conclusion

Several ticks and tick-borne diseases exist in the Caribbean, constituting sources of danger for human and animal health. Many of these pathogens are restricted to cycles of transmission primarily involving domestic Ehrlichia animals. canis) and wild animals and/or livestock (e.g. Ehrlichia ruminantium), while others are comparatively widespread in humans *Anaplasma mixtum, Anaplasma variegatum, and R. sanguineus* are the important vectors in the region due to their potential for transmission of zoonotic pathogens.

However, the Caribbean remains largely understudied with respect to tick-borne diseases because existing studies focus on the identification and epidemiology of pathogens affecting animals such as *E. ruminantium, Babesia (bovis and bigemina), and A. marginale*. But they were carried out using conventional detection tools which proved to be far from being able to answer questions concerning the risk of emergence or re-emergence of tick-borne diseases.

The use of the surveillance method based on the sequencing of new generations of microorganisms and pathogens present in ticks in Guadeloupe and Martinique undoubtedly represents a major improvement in epidemiological monitoring techniques which will allow the rapid and parallel detection of a wide range of pathogens. The ideal would be to be able to apply this method to the high-throughput screening of infectious agents present in ticks collected throughout the Caribbean, at least at the level of all the member countries of the Caribbean Animal Health and Veterinary Public Health Network (CaribVet).

The situation of the Republic of Haiti requires, in this regard, a little more attention because the majority of studies that have been carried out on ticks and tick-borne diseases have been carried out by foreign researchers. There is a great lack of professionals specialized in veterinary and medical entomology in this country, and the "One Health" approach is still slow to establish itself there because of a certain compartmentalization between the sectors of health, agriculture, and the environment. At government level, there does not yet seem to be much importance given to vector-borne diseases, with the exception of those transmitted by certain species of mosquitoes. The research undertaken with the veterinary parasitology and microbiology and public health laboratories did not allow us to discover any interesting data likely to advance the state of knowledge on this topic.

Research on ticks and tick-borne diseases must therefore be approached through networks of researchers throughout the Caribbean to facilitate exchanges of experience. The Caribbean Network for Animal Health and Veterinary Public Health (CaribVet) can therefore play a major role in the coordination of research and training actions on this theme.

Author details

Max Francois Millien*, Daphenide Saint-Louis and Daphnée Michel

Laboratoire de recherche sur les zoonoses et intoxications alimentaires (LAREZIA), Université Quisqueya, Haiti

*Address all correspondence to: maxfrancoismillien@gmail.com

References

[1] Jones KE, Patel NG, Levy MA, Storegard A, Balk D, Gittleman JL, et al. Global trends in emerging infectious diseases. Nature. 2008;**451**:990-993

[2] Moutailler S, George JC, Hansmann Y, Degeilh B, Joncour G, Jourdain E, et al. Principales maladies transmises par les tiques : épidémiologie, clinique et diagnostic. Tiques et maladies à tiques. 2015;**2015**:193-209

[3] Gondard M, Cabezas-Cruz A, Charles RA, Vayssier-Taussat M, Albina E, Moutailler S. Ticks and tick-borne pathogens of the Caribbean: Current understanding and future directions for more comprehensive surveillance. Frontiers in Cellular and Infection Microbiology. 2017;7:490

[4] Bonnet S, Huber K, Joncour G, René Martellet M, Stachurski F, Zenner L. Biologie des tiques. Tiques et maladies à tiques. 2015;**2015**:53-84

[5] Mans BJ, Gothe R, Neitz AVH. Biochemical perspectives on paralysis and other forms of toxicoses caused by ticks. Parasitology, Cambridge University Press. 2004;**129**(S1):SS95-S111. DOI: 10.1017/S0031182003004670[Opens in a new window]

[6] Guglielmone AA, Robbins RG, Apanaskevich D, Petney TN. The Argasidae, Ixodidae and Nuttalliellidae (Acari: Ixodida) of the world: a list of valid species names. 2010.

[7] Jongejan F, Uilenberg G. The global importance of ticks. Parasitology. 2004;**129**:S3-S14. DOI: 10.1017/s0031182004005967

[8] Mandell D. Principles and Practice of Infectious Diseases. 9th Edition. 2019;**2**. ISBN-13: 978-0323482554, ISBN-10: 0323482554

[9] Rhodain F, Perez C. Précis d'Entomologie Médicale et Vétérinaire. Paris: Maloine; 1985. p. 458

[10] Centers for Disease Cpntrol and Prevention (CDC). Ticks [Amblyomma spp.] [Dermacentor spp.] [Ixodes spp.] [Ornithodoros spp.] [Rhipicephalus spp]. 2017

[11] Waladde SM, Rice MJ. Chapter 3- The sensory basis of tick feeding behaviour. Physiology of Ticks. Current Themes in Tropical Science. 1982. pp. 71-118

[12] Franke J, Fritzsch J, Tomaso H, Straube E, Dorn W, Hildebrandt A. Coexistence d'agents pathogènes chez les tiques à la recherche d'un hôte et se nourrissant dans un seul habitat naturel en Allemagne centrale. Applied and Environmental Microbiology. 2010;**76**(20):6829-6836

[13] WHO. Vector-borne Diseases, Newsletter WHO; 2020

[14] McCoy KD, Chevillon C. Tiques et maladies à tiques - IRD Éditions – Open Edition Books Chapitre 4. Structuration des populations et adaptation des tiques : implications en épidémiologie. 2015

[15] Chauvin A, Moreau E, Bonnet S, Plantard O, Malandrin L. Babesia and its hosts: adaptation to long-lasting interactions as a way to achieve efficient transmission. 2008. Accepted: 16 April 2009

[16] Camus et Barré. Vector situation of tick-borne diseases in the Caribbean Islands. 1995. https://doi.org/10.1016/0303-4017(94)03118-G

[17] Agoulon A, Butet A, Hoch T, Perez G, Plantard O, Verheyden H, Vourc'h G. IRD Éditions Chapitre Dynamique des populations de tiques et liaison avec les facteurs environnementaux, 2015. pp. 85-112

[18] Chevillon C, Huber K. Tiques et maladies à tiques - IRD Éditions, 2015, pp. 141-164

[19] Nuttal et Labuda. Tick-host interactions: Saliva-activated transmission. Parasitology. 2004;**129**(7):S177-S189

[20] Belli A, Sarr A, Rais O, Rego RO, Voordouw MJ. Ticks infected via co-feeding transmission can transmit Lyme borreliosis to vertebrate hosts. Scientific Reports. 2017;**7**(1):1-13

[21] Starkey LA, Newton K, Brunker J, Crowdis K, Edourad EJP, Meneus P, et al. Prevalence of vector-borne pathogens in dogs from Haiti. Vet. Parasitology. 2016;**224**:7-12

[22] Rodriguez ON, Espaine L, Rivas A, Rodriguez P. Epidemiology of cattle diseases caused by haemoparasites in Cuba. 1989. Available at: http://www.cabi.org/isc/abstract/19902200923 [Accessed March 25, 2017]

[23] Goodell M, Thomas A, Rando TA. Stem cells and healthy aging. Science. 2015;**350**(6265):1199-1204

[24] Morel PC. Étude sur les tiques du bétail en Guadeloupe et Martinique. I. Les tiques et leur distribution (Acariens, Ixodoidea), 1966

[25] De la Cruz JO. Biogeography of the West Indies: Patterns and Perspectives. 2nd ed. Woods CA, Sergile FE, editors. Edition: 2. Illustrated, revised Publisher: CRC Press; 2001

[26] Basu A, Charles R. Ticks of Trinidad and Tobago: An Overview. 1st ed. Elsevier; 2017. Paperback ISBN: 9780128097441. eBook ISBN: 9780128097519

[27] Parola P, Vestris G, Martinez D, Brochier B, Roux VE, Raoult D. Tick-borne rickettiosis in Guadeloupe, the French West Indies: isolation of Rickettsia africae from Amblyomma variegatum ticks and serosurvey in humans, cattle, and goats. The American Journal of Tropical Medicine and Hygiene. 1999;**60**(6):888-893

[28] Dantas-Torres F. Biology and ecology of the brown dog tick, hipicephalus sanguineus. Parasites & Vectors. 2010;**3**(1):1-11

[29] Dantas-Torres F, Chomel BB, Otranto D. Ticks and tick-borne diseases: A One Health perspective. Trends in Parasitology. 2012;**28**:437-446

[30] Baneth G. Tick-borne infections of animals and humans: a common ground. International Journal for Parasitology. 2014;**44**(9):591-596

[31] Charles RA, Bermúdez S, Banović P, Alvarez DO, Díaz-Sánchez AA, Corona-González B, et al. Ticks and tick-borne diseases in Central America and the Caribbean: A one health Perspective. Pathogens. 2021;**10**(10):1273

[32] Walker AR, Bouattour A, Camicas J-L, Estrada-Peña A, Horak I, Latif A, et al. Ticks of Domestic Animals in Africa: A Guide to Identification of Species. University of Edinburgh; 2003. Available online: https://agris.fao.org/agris-search/search.do?recordID=XF2015039701

[33] Guglielmone AA, Beati L, Barros-Battesti DM, Labruna MB, Nava S, Venzal JM, et al. Ticks (Ixodidae) on humans in South America. Experimental & Applied Acarology. 2006;**40**(2):83-100

[34] De Castro JJ. Sustainable tick and tickborne disease control in livestock improvement in developing countries. Vet. Parasitology. 1997;**71**:77-97

[35] Uilenberg G, Zivković D, Dwinger RH, Ter Huurne AA, Perie NM. Cross immunity between strains of Cowdria ruminantium. Research in Veterinary Science. 1983;**35**(2):200-205

[36] Camus et Barré. The role of Amblyomma variegatum in the transmission of heart water. 1992

[37] Loftis AD, Kelly PJ, Freeman MD, Fitzharris S, Beeler-Marfisi J, Wang C. Tick-borne pathogens and disease in dogs on St. Kitts, West Indies. Veterinary Parasitology. 2013;**196**:44-49

[38] Pegram R, Indar L, Eddi C, George J. The Caribbean Amblyomma program some ecologic factors affecting its success. Annals New York Academy Sciences. 2004, 2004;**1026**:302

[39] Scoles GA, Ueti MW. Amblyomma cajennense is an intrastadial biological vector of Theileria equi. Parasites & Vectors. 2013;**6**(1):1-9

[40] Asgarali Z, Coombs DK, Mohammed F, Campbell MD, Caesar E. (2007). A serological study of Babesia caballi and Theileria equi in Thoroughbreds in Trinidad. Vet. Parasitology. 2007;**144**:167-171

[41] Georges K, Ezeokoli CD, Newaj-Fyzul A, Campbell M, Mootoo N, Mutani A, et al. The application of PCR and reverse line blot hybridization to detect arthropod-borne hemopathogens of dogs and cats in Trinidad. Annals New York Academy Sciences. 2008, 2008;**1149**:196

[42] Rojas A, Rojas D, Montenegro V, Gutiérrez R, Yasur-Landau D, Baneth G. Vector-borne pathogens in dogs from Costa Rica: First molecular description of Babesia vogeli and Hepatozoon canis infections with a high prevalence of monocytic ehrlichiosis and the manifestations of co-infection. Veterinary Parasitology. 2014;**199**:121-128

[43] Díaz-Sánchez AA, Hofmann-Lehmann R, Meli ML, Roblejo-Arias L, Fonseca-Rodríguez O, Castillo AP, et al. Molecular detection and characterization of Hepatozoon canis in stray dogs from Cuba. Parasitology International. 2021, 2021;**80**:102200

[44] Sant C, Georges KC, Pow-Brown P. Novel incidental finding of Hepatozoon canis infection in two dogs of the same household in Trinidad, West Indies. Veterinary Parasitology: Regional Studies and Reports. 2017;**9**:98-103

[45] Guglielmone AA, Nava S, Robbins RG. Neotropical Hard Ticks (Acari: Ixodida: Ixodidae). In: Englischer Sprache Springer. 1st Edition. Springer International Publishing; 2021. ISBN 978-3-030-72352-1

[46] Kelly PJ, Köster L, Li J, Zhang J, Huang K, Branford GC, et al. Survey of vector-borne agents in feral cats and first report of Babesia gibsoni in cats on St Kitts, West Indies. BMC Veterinary Research. 2017;**13**:331

[47] Kelly P, Lucas H, Beati L, Yowell C, Mahan S, Dame J. Rickettsia africae in Amblyomma variegatum and domestic ruminants on eight Caribbean islands. J. Parasitology. 2010;**96**:1086-1088

[48] Kelly PJ. Rickettsia africae in the West Indies. Emerging Infectious Diseases. 2006, 2006;**12**:224

[49] Aitken TH, Jonkers AH, Tikasingh ES, Worth CB. Hughes virus from Trinidadian ticks and terns. Journal of Medical Entomology. 1968;**5**(4):501-503

[50] Jonkers AH, Casals J, Aitken THG, Spence L. Soldado virus, a new agent from Trinidadian Ornithodoros ticks. Journal of Medical Entomology. 1973;**10**(5):517-519

[51] Danielová V, Marhoul Z, Dusbábek F, Ryba J, Fernández A, De La Cruz J, et al. Isolation of Hughes virus from ticks in Cuba. Acta Virologica. 1982;**26**(3):186-189

[52] Málková D, Holubová J, Cerný JV, Daniel M, Fernández A, de la Cruz J, et al. Estero real virus: a new virus isolated from argasid ticks Ornithodoros tadaridae in Cuba. Acta Virology. 1985;**29**(3):247-250

[53] Belle EA, King SD, Griffiths BB, Grant LS. Epidemiological investigation for arboviruses in Jamaica, West Indies. American Journal of Tropical Medicine. 1980;**29**:667-675

[54] Butler JF, Gibbs EPJ. Distribution of potential soft tick vectors of African swine fever in the Caribbean region (Acari: Argasidae). Preventive Veterinary Medicine. 1984;**2**(1-4):63-70

[55] Hess WR, Endris RG, Haslett TM, Monahan MJ, McCoy JP. Potential arthropod vectors of African swine fever virus in North America and the Caribbean basin. Vet. Parasitology. 1987;**26**:145-155

[56] Endris RG, Haslett TM, Hess WR. Experimental transmission of African swine fever virus by the tick Ornithodoros (Alectorobius) puertoricensis (Acari: Argasidae). Journal of Medicine Entomologiste. 1991;**28**:854-858

[57] Pérez-Eid C. Les tiques: identification, biologie, importance. Lavoisier: Google Books Publisher; 2007

[58] Kazimírová M, Štibrániová I. Tick salivary compounds: their role in modulation of host defenses and pathogen transmission. 2013

[59] Rodríguez-Mallon A, Fernández E, Encinosa PE, Bello Y, Méndez-Pérez L, Ruiz LC, et al. A novel tick antigen shows high vaccine efficacy against the dog tick, Rhipicephalus sanguineus. Vaccine. 2012;**30**(10):1782-1789

[60] Wood CL, Lafferty KD. Biodiversity and disease: a synthesis of ecological perspectives on Lyme disease transmission. Trends in Ecology & Evolution. 2013;**28**(4):239-247

[61] Zhang J, Kelly P, Guo W, Xu C, Wei L, Jongejan F, et al. Molecular detection of Theileria spp. in Livestock on Five Caribbean Islands. BioMedical Research International. 2015;**2015**:e624728

[62] Obregón D, Cabezas-Cruz A, Armas Y, Silva JB, Fonseca AH, André MR, et al. High co-infection rates of Babesia bovis, Babesia bigemina, and Anaplasma marginale in water buffalo in Western Cuba. Parasitology Research. 2019;**118**(3):955-967

Chapter 9

Arthropods: Prospect of Household Food Security

Jonathan Ibrahim and Dalyop Daniel Gyang

Abstract

Food security is a "situation that exists when all people, at all times, have physical, social, and economic access to sufficient, safe, and nutritious food that meets their dietary needs and food preference for an active and healthy life". With a growing world population and increasingly demanding consumers, the production of sufficient protein from livestock, poultry, and fish represents a serious challenge for the future and prompts the need for other sources of nutrition to be explored. Approximately more than 1,900 arthropod species are edible. This requires the development of cost-effective, automated mass-rearing facilities that provide a reliable, stable, and safe product for consumption. This chapter discusses arthropods as food, arthropods as animal feed, nutritional composition, the secondary metabolites of edible insects and potential medicinal substances, development and utilization of edible insect's resources, insect farming, impact of insect quality on consumers' preference and acceptability (insect processing and product quality, processing and marketing, and consumer acceptance), food safety and legislation, as well as the way forward.

Keywords: arthropod, prospect, household, security, impact

1. Introduction

The largest phylum in the animal kingdom, which includes well-known insects, spiders, ticks, and crustaceans, as well as numerous smaller, lesser-known species and a plethora of bizarre forms only known as fossils. Arthropods account for approximately 95% of all animal species. The number of recognized species is expected to be in excess of one million, with insects accounting for the majority. Nobody is sure how many arthropod species there are. Authorities believe it might be in the tens of millions [1]. The body of an adult arthropod is normally made up of a succession of ring-like segments with a pair of numerous jointed limbs on each segment that move on each other via muscles. However, other parasites, such as pentastomids and rhizocephalans, show no signs of segmentation as adults. Arthropods' integument produces a stratified cuticle containing chitin. This exoskeleton must be shed regularly to allow for growth, a process known as molting or ecdysis. Young stages change significantly from adults, and some parasitic organisms have extremely distinct body forms than their closest relatives. Arthropods are distinguished from all other creatures by their traits [1].

Food insecurity is worsening as a result of rising population and constraints on food importation, among other things. As a result, there is a high prevalence of hunger and malnutrition, with children and women being particularly susceptible. Apart from the increased risk of hunger brought on by the deteriorating food situation, the widespread prevalence of Protein Energy Malnutrition (PEM) has resulted in high rates of illness and death, particularly among babies and children in developing nations. While every effort is being made to increase food production through conventional agriculture, including current interest in the possibilities of exploring the vast number of less familiar plant resources that exist around the world [2], almost no attention has been paid to the consumption of Arthropods, a traditionally recognized and available source of protein and fats. Furthermore, protein meals are scarce, making them out of reach for low-income households, which sadly make up the majority of the population in most emerging nations [3]. The scarcity of common animal nutrition sources and the high expense of the few available plant sources should spur urgent study into the nutritious potentials of arthropods.

2. Arthropods as food

Most species of arthropods are obtained from nature in tropical nations. More than 2000 bug species are found in an inventory of edible insect species eaten across the world using solely scientific names rather than common ones. In terms of most species consumed, certain countries stand out. This, on the other hand, is primarily due to the amount of research completed. Ramos-Elorduy, for example, published a large number of publications in Mexico on entomophagy (the eating of insects), and Belgian scientists discovered more than 60 edible caterpillars in the Democratic Republic of Congo, a former Belgian colony [4]. This also implies that many edible bug species are yet unknown, necessitating more research [4]. Insects are more commonly consumed in tropical regions because they are bigger and generally congregate in clumps, making gathering easier. In addition, because there is no winter season, bug species can be found throughout the year. Most insect species are seasonal because they are dependent on the availability of their host plant; others, such as most aquatic insects, can be found all year. Beetles (31%), caterpillars (18%), wasps, bees, and ants (15%), crickets, grasshoppers, and locusts (13%), true bugs (11%), and termites, dragonflies, flies, and others (12%) are among the insects consumed [4, 5].

Spiders and scorpions are examples of arthropods that are eaten. Some species are semi-domesticated, meaning that some precautions are taken to make harvesting more predictable [6]. Palm trees, for example, may be chopped down to encourage palm weevils of the genus Rhynchophorus (*Coleoptera: Curculionidae*) to lay their eggs on the trunk. The larvae are gathered when a particular amount of time has passed. In many areas of the world, these larvae are considered a delicacy. In the tropics, nothing is known about how frequently and how much insects are consumed [7]. This is due to the fact that insects are not counted as food or feed in national agricultural statistics. In underdeveloped nations, the vast majority of insects are collected from natural populations in nature, farmlands, or woods. Edible insects provide a low-cost and effective way for vulnerable groups to enhance their livelihoods and the quality of their traditional foods. In western nations, the use of insects as food has recently acquired popularity. Several businesses have begun to breed insects for human consumption. In the United States, for example, crickets are frequently used in processed foods like as protein bars. They are already available in supermarkets in several countries.

3. Arthropods as animal feed

Arthropods have been studied as a feed constituents for aquatic and domestic animals. Insect meal has been shown to have appropriate palatability for chickens, pigs, fish species, and ruminants. Insects can replace 25–100% of soymeal or fishmeal depending on the animal type [8]. The Black soldier fly *Hermetia illucens (Diptera: Stratiomyidae*) and the Domestic house fly *Musca domestica* are the most promising species for large-scale production *(Diptera: Muscidae)*. Mealworms, termites, grasshoppers, crickets, and caterpillars are among the other species that are evaluated (such as the silkworm). The use of the Black soldier fly as a feed for chickens, pigs, channel catfish, African catfish, blue tilapia, turbot, and rainbow trout has been examined [8]. Agricultural by-products such as coffee pulp, palm kernel meal, and manure, as well as organic waste materials such as fish offal, market waste, municipal organic waste, dewatered fecal sludge, organic leachates, and Distiller's Dried Grains with Solubles (DDGS) can all be recycled using fly larvae [9].

The use of insects in aquaculture has recently attracted a lot of attention. This is due to the decreasing availability of fishmeal as a primary source of dietary protein in compounded feed for a number of important farmed species [10]. Fishmeal is manufactured from pelagic fish caught in international seas. International fisheries are overfished, and existing techniques are unsustainable [11]. Fish and shellfish farming has been the fastest growing food producing sector in the previous several decades (it is still expanding at 6% per year) and has become a major business in many nations, increasing demand for fishmeal [11]. In 2012, farmed food fish accounted for 42% of all fish produced worldwide, including both capture and aquaculture (it was just 13% in 1990) [11]. As a result, fishmeal and fish oil output has decreased from 30 million tons (live weight) in 1994 to 16 million tons in 2012 [12]. This scarcity has driven a hunt for other protein sources, including the utilization of insects [11]. Insect meal is a promising alternative to soymeal in aquaculture, as vegetable-based diets have a number of drawbacks. These include amino acid imbalances, anti-nutritional elements, low palatability, and a large amount of fiber and non-starch polysaccharides [13].

Insects have been allowed in aquaculture feed since 2013, according to a European Union (EU) regulation. In Norway, research has shown that insect meal is an excellent protein source for farmed salmon [14]. The Norwegian Research Council has invested over one million euros to research the possibilities of employing insects as a safe and nutritious fish feed element [14].

4. Nutritional composition

The nutritional profile of edible insects is difficult to generalize. Despite the wide variety, data from 236 edible insect species demonstrate that they offer adequate energy and protein, fulfill human amino acid needs, are high in monounsaturated and polyunsaturated fatty acids, and are rich in various minerals and vitamins [15, 16]. In comparison to normal beef, the high iron and zinc concentration is particularly interesting. As a result, entomophagy has been recommended to address these mineral deficiencies in underdeveloped nations, particularly in light of the fact that the global population at risk for these deficiencies is more than 17% for zinc and 25% for iron [17].

Mealworms and crickets, for example, contain a protein level of 19 to 22%, making them suitable for human consumption [18]. In terms of protein content, this is equivalent to typical meat products [17]. The scientists found that the necessary amino acid levels in the insect species they studied were equivalent to soybean proteins, but lower than casein. Gels might be produced and soluble fractions obtained using a simple water extraction process for potential culinary applications [19].

4.1 Edible insect protein

Insects can be a more effective source of protein, and edible insects have a bright future [20]. Protein concentration varies depending on the insect condition, according to studies. Adults have the largest protein concentration, followed by pupas, and finally larvae [20]. According to protein estimates of insects of various ages, the adult has 71.07% protein content, the pupa has 58.59% protein content, and the larva has 50.83% protein content [5]. The protein composition of insects from various tropics differs as well. Orthoptera is ranked higher than Homoptera, as well as Odonata, Diptera, Hymenoptera, Hemiptera, Lepidoptera, and Coleoptera [21].

Amino acid is the fundamental functional unit of biological macromolecular protein, as well as a significant component of the food that insects consume. The amino acid concentration of edible insects ranges from 10 to 70%, with 10 to 30% being necessary amino acids. The majority of insect amino acid ratios are adequate and have approached or even exceeded the WHO/FAO recommended ratios [5]. The presence of a considerable number of free amino acids linked to insect freshness was also discovered [22]. The content of free amino acid of edible insects in the blood is about 3000–23,400 mg/kg, which is more than any other higher animal in the cosmos [22].

4.2 Carbohydrate of edible insects

Edible insects' carbohydrates (sugars) are highly rich in glucose, triose, glycogen, erythritol, ketose sugar, fructose, and ketoheptose. The sugar content of sea algae (the constituent blood sugar of insects) is also high [23]. Edible insects carbohydrates are simple to digest and absorb with total sugar content ranging from 1–10% or even lower which is highly good for human health [24]. The total sugar content of *Cyclopelta parva* is 1.45%, while *Tessaratoma papillosa* has a sugar concentration of 0.15% [20].

The major component of edible insect skin and bones is chitin. N-Acetyl-D-glucosamine copolymer is its chemical name, and it has adsorption properties for a certain toxin. It is also a low-calorie food with a high nutritional content that is beneficial to one's health. Chitin aids in the prevention of high blood pressure by promoting intestinal peristalsis, weight reduction owing to fat, antiaging, enhancing immunological function, and so on. The edible insect body content of chitin is generally between 15 and 18%. Chitin content varies according to insect nature, such as the chitin content of dry silkworm pupa (3.73%) and Skim pupa's content is 5.55% [25].

4.3 Mineral elements and vitamin of edible insects

Edible insects are high in mineral elements such as calcium, phosphorus, iron, and zinc, among others, which are frequently required by the human body as supplements. Feed insects have been reported to be able to meet the Fe, Cu, Zn, and Mg mineral requirements of animals [26]. Mineral elements such as Mn, Fe, Cu, and Zn are found in abundance in locusts [26]. Zn, Se, Mn, and Mg [27] are abundant in many ants. Edible insects are high in Se, Co, Ni, and Cd trace elements, in addition to the constant element (73). The Se content of the Chinese rice locust and yellow powder bug is 4.62 and 4.75 mg/kg, respectively [28]. Se can help with detoxification, carcinogenic activity inhibition, carcinogen destruction, and cancer cell growth and division prevention [28]. Other elements found in *Formica (Coptoformica) mesasiatica Dlussky* [28] include Ni (1.22 mg/g), Co (1.36 mg/g), and Cr (1.52 mg/g).

Vitamins B1 (thiamine), B2 (riboflavin), B3 (niacin), B6 (pyridoxine), C, D, E, K, and carotene are all found in the bodies of insects [28]. *Macrotermes annandalei* contains 25.0 I.U./g of vitamin A, 85.4 I.U./g of vitamin D, and 11.7 I.U./g of vitamin E. Vitamins are necessary for sustaining the human body's regular physiological function [28].

4.4 Lipid contents of edible insects

Oil and fats are abundant in insects [20]. Pupae and larvae have a greater fat content than adult insects [20]. The fat content of the insect decreases once it has feathered. The fat content of edible insects is usually between 10 and 50% [20]. Unsaturated fatty acid and palmitic acid are higher in edible insects [20]. Among them, linolenic acid content is higher. They can be employed as medicinal raw materials in the form of textiles and stencils [20]. The fat content of wasps was discovered in a recent study. Fat content in larvae is 29.01%, in pupae it is 27.25%, and in adults it is 17.22% [20]. Lepidoptera have more unsaturated fatty acids and palmitic acid, while *Coleoptera* have more oil acids [20]. Infrared spectroscopy of insect wax revealed that it is mostly made up of long-chain hydrocarbons, fatty alcohols, fatty acids, and some molecules with aromatic rings mixed in [29].

5. The nutritional evaluation of insect oil/fat

Insect oils (fat) are nutritional compounds with a wide range of physiological and biological effects. It is highly valued in research, development, and application, regardless of quantity or quality [30]. The fat content of an insect's body varies during its life cycle. It is intimately linked to the insect species' development [30]. Many studies have shown that the fat content of insects varies between species. In the same species, the oils (fat) of the pupa and larva were greater than those of the adults. The oil content of the insects was also greater throughout the winter [30].

The dry body fat level of insects was typically 10%, while many other insects have fat content of 30%, or even up to 77.16% [30]. Insects are high in fat and have a balanced fatty acid profile. The ratio of saturated to unsaturated fatty acids in edible insects is usually less than 0.4 [30]. Its partial fatty acid content ratio is similar to that of fish, therefore it may be utilized as a natural health care product. Insects' saturated fatty acids (SFA) are largely made up of palmitic acid (C16:0), not stearic acid (C18:0), which is abundant in vertebrates [30]. In addition, insect oil contains carbon fatty acids with an odd number of carbons, such as pentadecanoic and heptadecanoic acids, which are very unusual in nature but exceedingly frequent in insects [27]. The concentration of heptadecanoic acid in termite adults, housefly larvae, and housefly adults was all greater than 2% [27]. Because odd-number carbon fatty acids have a unique raw active action, they were discovered to have higher anticancer activity [27]. As a result, many researchers are interested in the enrichment and separation of odd number carbon fatty acids in insects, resulting in a hotspot in insect oil research. Insect oil is a natural active product solvent that contains lecithin and fat soluble D raw ingredient (such as vitamin A, D, E). These active natural compounds have a significant physiological and biological function with a high value [30].

6. The secondary metabolites of edible insects and potential medicinal substances

A great number of researches in recent years have demonstrated that insect secondary metabolites are valuable sources for discovering novel leading chemicals [31]. Compounds generated from fatty acid, polyketide, terpenoid, nucleoside, and amino acid routes are among the structurally varied arthropod natural products having insect constituents. However, majority of these chemicals' production has not been well investigated [31].

The historic use of plants as remedies, referred to as "ethnobotany," has long been acknowledged and investigated [29]. Insects have long been used as remedies in a variety of civilizations, particularly in traditional Chinese medicine. It might be beneficial in the creation of effective medications. Another ongoing project is finding novel antibacterial structures from natural insect products. More recent investigations [31] are being conducted to investigate the therapeutic effects of isolated chemical components from insects and other arthropods.

7. Development and utilization of edible insect's resources

Based on the diverse insects eaten resources classification, edible insects may be split into food insects, drug/medicinal insects, and drug dual-use insects, among others [22]. Edible insects are intended for everyday food intake, and they have significant nutritional value for humans to create and use. In 2013, the United Nations' Food and Agriculture Organization (FAO) published "Edible Insects: Future Prospects for Food and Feed Security" [22]. It explains the several advantages of consuming insects for humans all around the world. It was proposed in the Fifth Latin American Congress of Dietitians and Nutritionists in 1980 to replace the human food scarcity, which sees them as part of the food source. Insects as human food in many countries has become increasingly visible now [22]. Red ants, grasshoppers, and some predaceous diving beetles (Dytiscidae) have enough insect protein to compete with lean beef, according to scientists. Adult insect protein content is abundant, far exceeding that of pork, cattle, poultry, fish, and eggs. Insects will be the third group in the future, following cell raw material and microbial protein sources, according to experts [32]. People in disadvantaged areas require important nutrients, and the services of insects and spiders are equally beneficial. In industrialized countries such as the United States, insects and spiders are higher protein foods that are considered healthful. With high fat, protein, vitamin, fiber, and mineral content, insects are a very nutritious and beneficial dietary source [22].

At the home level or in bigger industrial scale enterprises, gathering and growing insects can provide employment and economic revenue. It has the potential to employ millions of people all around the world (Reference needed). Furthermore, data suggests that most breeding insects emit less detrimental greenhouse emissions to the environment than animals (Reference needed). This discovery will aid in the reduction of food production costs and greenhouse gas emissions. In China, the processing technology of the functional food and health food industries of edible insects has accelerated at an unprecedented rate in recent years, in tandem with the advancement of contemporary science and technology. For instance, a concentrated insect protein oral beverage including honey, royal jelly, pollen, and propolis, as well as the customary shellac ash. (Reference is required) Some insect oils are primarily employed as fat soluble functional components [32].

7.1 Insect farming

The majority of bug species in tropical regions are obtained from nature. Insects, on the other hand, must be cultivated like mini-livestock if they are to become a valuable resource. Furthermore, edible insect resources in nature are already under pressure from over-exploitation, habitat deterioration, and pesticide usage [33, 34]. The collecting and selling of the Mopane caterpillar *Imbrasia belina* (*Lepidoptera: Saturniidae*), for example, jeopardizes the long-term usage of forestry resources. As a result, a harvesting period restriction has been proposed [35].

Thailand is one of the countries where insect farming plays a vital role, with 20,000 farms producing roughly 7500 tons per year with operations spreading into Laos [36]. Several multinational programs are currently functioning in Africa to encourage insect breeding for human consumption, with a focus on crickets [37]. Farming commercially significant insect species like the Mopane caterpillar has been attempted [38]. However, virus transmission within a confined population is still an issue, and it is not yet economically feasible [38]. Bug rearing firms in the Western world manufacture a variety of insect species for pet food [39]. Some firms in the Netherlands have set up dedicated manufacturing lines for human consumption of mealworms, crickets, and locusts [39]. When insects are used as feed, however, feedstock firms demand big, consistent, and consistent supply, which can only be produced in industrial automated raising facilities [39]. Increased insect production for food and feed on a big scale will provide numerous new hurdles, including disease issues. The *Acheta domesticus densovirus* (AdDNV) is one example, which has decimated commercial house cricket (*Orthoptera: Gryllidae*) rearing throughout Europe and portions of North America [40].

Nigerians gather edible insects from the wild, which is hampered by seasonality, quality time wasted during collecting, and little quantity obtained (Reference needed). As a result of its scarcity, the supply of edible insects will be disrupted, resulting in high cost. In Nigeria, insect farming will help with food provision, particularly in rural areas that are closer to the wild additional revenue for essential expenses such as food, agricultural equipment, and education; and it will help with food shortages due to seasonal drought in output (Reference needed). Insect farming for food also provides chances for landless people and women involved in the collecting, farming/cultivation, processing, and sale of insects to improve their nutrition, get employment, and earn money [41, 42]. Edible insects may be found in all of Nigeria's ecological zones, with each geopolitical zone having a unique edible bug that can be grown for profit. According to Alamu *et al.* [43], of the 22 most eaten insects in Nigeria, 77.3% were Lepidoptera (27.3%), Coleoptera (27.3%), Orthoptera (22.7%), and Isoptera, Hemiptera, and Hymenoptera (22.7%). This demonstrates that edible insect consumption in Nigeria is diverse. Only a small number of insects are farmed in Nigeria, and those that are farmed are not in commercial quantities. *Rhynchophorus spp*., or palm weevils, are excellent low-cost providers of vital nutrients. They are extremely tasty and are commonly prepared roasted, and they have a low carbon footprint when farmed commercially. Palm weevils are a traditional meal for most rural societies (especially in the south), but they are not farmed for consumption; instead, they are harvested in the wild. Muafor *et al.* [44] described traditional grub harvesting and grub semi-farming as indigenous ways of palm weevil cultivation. These agricultural techniques account for 30 to 75% of household income. Ebenebe & Okpoko [45] reported that, palm weevils were reared on eight distinct culturing substrates (coconut fiber, coconut fiber with palm wine, mahogany sawdust, mahogany sawdust with palm wine, palm frond petiole, palm bunch midrib, sugarcane tops (SCT), and spoilt water melon (SWM)). In terms of materials and labor, palm weevil farming is a cost-effective business. Within three to four months, the larvae attain adulthood and may be collected for eating; they are high in protein.

Cirina forda is a Nigerian delicacy eaten mostly by people from the south. It has a high protein, fat, and necessary mineral content. Because there is now no commercial farm providing this wonderful protein source, cultivating it for food will be profitable. A large number of the insects are collected in the wild before being processed and sold in major marketplaces in Nigeria's southwest. *Cirina forda* larvae may be grown on the leaves of a growing *Vittelaria paradoxa* tree, according to Ande and Fasoranti [46]. All instars matured and were gathered within one month. Despite the huge potential contained in the larvae, they have not been economically produced in decades. Starting this business has the potential to be very profitable. According to Ebenebe and Okpoko [45], cricket (*Gymnogryllus lucens*) is the most popular insect eaten in Nigeria, but it can only be caught in the wild.

8. Impact of insect quality on consumers' preference and acceptability

Nutrient content (protein being a key component), insect quality (especially taste, flavor, look, palatability), and external variables (availability, easy pricing, suitable social milieu) are all essential considerations in accepting insects as food [47]. Forest-dwelling people have easy access to natural regions, not only in underdeveloped nations but also in rural areas such as Japan [34]. Consumer reactions to wild insects and their food products, as well as their preferences, acceptability, and consumption of insect-based meals, are currently unknown. Anecdotal evidence suggests that in Africa and India, certain wild edible insect species are preferred and accepted above cultivated ones such as silkworms or crickets [34].

Alemu *et al.* [43] observed no significant difference in whole or powdered termite eating in Kenya. Before purchasing termites, such as *Macrotermes falciger*, buyers evaluated the insect stock for freshness, presence of legs, cleanliness, species type, and oil content at the local market. Fried adults were selected by the majority of purchasers (77.6%) [43]. Long-bodied termite soldiers were in high demand and favored over late variants [48]. In Tanzania, Kenya, and Uganda, the grasshopper (*R. differens*) is a traditional delicacy, a source of nutrition, and a delightful multi-functional insect [47]. Consumers preferred salted, boiled, and smoked grasshoppers or deep fried grasshoppers in cotton seed oil above any other single processing technique (smoking, deep frying, sun-drying, toasting, boiling) [47]. *R. differens* adults that had been cooked with salt, onion, and tomato and then dried were favored above those that had merely been deep-dried with salt and onion in another study. The acceptability ratings for these goods were 7.2 and 5.2, respectively (on a scale of 0–9, with 9 being the highest approval) [48].

People in Uganda favored boiled and dried grasshoppers with salt, onion, and tomatoes to those that were just boiled and dried without tomatoes in the case of *R. nitidula* [48]. With the exception of grasshoppers, whose legs are often removed, entire insects are valued in India; larvae, pupae, and adult termites are sometimes combined and sold together by local sellers [49]. Overall, while there was a higher acceptance for insects without much attention to the species, fear of trying an unknown product, lack of taste experience, and a belief of low social acceptance were identified as major barriers to popularizing edible insects [50]. Despite the fact that taste alone failed to distinguish insects from cheese or bread in more than half of the probands tested there was a very low acceptability [51]. Acceptability is also influenced by accurate labeling. Siozios, [52] discovered many discrepancies in identification in packets containing mopane caterpillars, winged termites, and grasshoppers while testing the correctness of insect goods on the UK market. This may make customers less willing to accept and consume insects or insect products.

Information on entomophagy, past experience and familiarity with edible insects, look, flavor, and overall likability of a species are all important considerations when choosing edible species. As a result, views regarding insects as food and food supplemented with edible insects can be influenced by information and knowledge [53]. In fact, a survey conducted two years following the introduction of edible insects in Belgium demonstrated a rising favorable reaction in terms of acceptability, according to Van Thielen *et al.* [54]. A comparable poll of Danish customers found that 23% of them would eat insects [55].

8.1 Insect processing and product quality

Traditionally, insects are consumed raw or processed (dried, crushed, pulverized, grounded, pickled, cooked, boiled, fried, roasted/grilled, toasted, smoked or extruded [54]. Besides these techniques, Kewuyemi *et al.* [56] suggested fermentation to enrich the inherent composition of insect-based products and to induce anti-microbial, nutritional and therapeutic properties. Similarly, defatted *T. molitor* larvae and oil could be used as food ingredients. Defatted mealworm powder is high in protein, minerals, and bioactive substances, and has a savory flavor due to the abundance of amino acids. The oil is high in tocopherol and has a long shelf life [57]. Insects are frequently fasted before processing, and big specimens are degutted or defatted since the gut may contain undigested plant material, excreta, bacteria, and other contaminants; also, degutted insects have greater crude fiber protein levels [56]. Tribal communities have consistently embraced this approach since it is efficient and practical, especially for huge lepidopteran larvae. Insects that have been processed can be freeze-dried, sun-dried, or canned. Consumer preferences, insect species availability and compatibility, social custom, religious rites, tribal ethics, and family tradition may all influence processing procedures [58]. Anuduang *et al.* [59] investigated the antioxidant characteristics of silkworm powder at four different drying temperatures (80, 100, 120, and 140°C) and found that the lowest drying temperature maintained the most phenolic compounds and antioxidants.

When choosing a food item based on "post-ingestive fitness," the processing method can aid in the removal of anti-nutrients and other harmful components while also extending the shelf life. As a result, processing is necessary to retain nutritional content, increase shelf life, and obtain functional and fortified foods

[57]. Products are supplemented with insect chitosan (a polysaccharide derivative of chitin) in food processing facilities, which is more soluble and so favored over raw chitin [60]. Traditional wisdom based on centuries of experience is regularly used by local communities to improve insect-based cuisine [61]. Methods can, of course, evolve and be replaced by others, since each has benefits and disadvantages that are tailored to area circumstances. In North-East India, for example, roasting, grilling, and frying are commonly used because insects taste better than boiling and baking [62].

Vitamins are generally heat sensitive, and heat processing reduces the quantity of these essential chemicals fully or partially. To avoid insect damage, storage conditions are critical. However, whereas the level of tocopherol in *T. molitor* and *Zophobas morio* did not change in various settings [63], the antioxidant capabilities of silkworm powder did. Nyangena *et al.* [64] investigated the effects of traditional processing techniques on the proximate composition and microbiological quality of *Acheta domesticus, Ruspolia differens, Hermetia illucens*, and *Spodoptera litoralis*, including boiling, toasting, solar-drying, oven drying, boiling + oven drying, boiling + solar-drying, toasting + oven-drying, toasting + solar-drying, toasting + oven-drying. Traditional processing enhanced microbiological safety but reduced nutritional value, according to the researchers [63].

8.2 Processing and marketing

The Mopane caterpillar trade is significant business in southern Africa. Styles estimated in 1994 that an annual population of 9500 million mopane caterpillars in South Africa's 20,000 km2 of mopane veld was worth more than US$ 80 million, with around 40% going to producers who are mostly impoverished rural women [59]. To alleviate child malnutrition, supplemental diets based on edible termites were designed and assessed in Kenya. It can be processed into economical and safe meals with acceptable nutrient density, according to the findings [65]. To stimulate entomophagy in Kenya, termites and lake flies were baked, boiled, and cooked to extend shelf life and then processed into common consumer items like crackers, muffins, sausages, and meat loaf. These are ways for making naturally gathered items available for extended periods of time. Techniques including drying, acidifying, and lactic fermentation can be used to preserve edible insects and insect products without the need of a refrigerator [30]. However, insects should be farmed to better manage and ensure the supply of such insect goods. Then freeze-drying (dehydration of the frozen insect via sublimation) is commonly used [30].

8.3 Consumer acceptance

In tropical areas, insects are a major source of protein, yet Europeans are wary of eating them. Bequaert ascribed western people's reluctance to eating insects to 'prejudice' and cultural, conditioning over a century ago: "What we eat and what we do not eat is, after all, a question of tradition and fashion (rather than) anything else" [66]. DeFoliart also saw the western mindset and prejudice against eating insects as a significant impediment to the introduction of this sustainable food source. Yen [67] predicted that 'westernization' of insect-eating communities would lead to a shift away from entomophagy, while western countries, as main consumers of cattle protein, would miss out on a chance to lessen their environmental impact. Others have emphasized the necessity for techniques to overcome the psychological and cultural hurdles to entomophagy, citing the value of insects as human food as a difficult test case [68]. In Belgium and the Netherlands, research found that motives for sustainable food consumption promoted the acceptance of insects as a protein source [67, 69]. Customers in Thailand, where insects are a part of the local culinary culture, saw insects differently from Dutch consumers in terms of flavor and familiarity. Mealworms, for example, were greatly disliked by Thai participants owing to their link with larvae found in decaying debris [67]. The Dutch individuals, who were more familiar with mealworms as food, did not have this link. The following solutions have been offered to alleviate the aversion to eating insects [67].

1. To raise customer awareness of the product by offering information about insects as a sustainable alternative food source.

2. To make edible insects available and to teach people how to cook them.

3. To emphasize the close relationship between insects and crustaceans in animal categorization.

4. To increase the number of edible bug exposures and taste tests.

5. Create appropriate items that not only lessen the barriers to trying new things, but also taste good and are enjoyable to consume.

6. Incorporating insects into everyday foods.

7. To employ role models such as Kofi Annan, the former UN Secretary-General, who was interviewed about edible insects.

8. To target youngsters for entomophagy instruction.

9. Food safety and legislation

A number of writers have addressed food safety concerns, with urgent legal implications. Contaminants such as heavy metals, mycotoxins, pesticide residues, and infections are all potential risks. The existing research on insect eating in tropical regions shows that insects gathered for human consumption do not pose any substantial health risks [70], but there is little information on insects cultivated for food or feed. Nobody regarded insects to be food or feed at the time the legislation was enacted, therefore when the word "animal" is used in the statute, insects are frequently included. The EU Regulation 1099/2009, for example, states that animals must be murdered in approved slaughterhouses in the presence of an Animal Welfare Officer; this plainly does not apply to insects [24].

In the case of insects as food, the EU has yet to rule whether an insect product is regarded a novel food because it was not consumed "in a considerable degree" in the EU prior to May 15, 1997. If this is the case, the manufacturer must supply a Novel Food Dossier, among other documents, demonstrating that the product is safe for consumers. The EU adopted a rule in 2013 permitting the use of non-ruminant proteins in aquaculture feed for fish; a removal of the restriction on insect proteins in feed for food-producing pigs and poultry is being studied [23].

Food containing insects, such as the Yellow mealworm *Tenebrio molitor* (*Coleoptera: Tenebrionidae*), may cause allergy reactions in those sensitive to home dust mites and crustaceans [25]. Recent evidence reveals that insects and crustaceans (such as shrimps), which have long been thought to be taxonomically distinct branches of the arthropod family tree, are really taxonomically linked [28]. When an insect product is found to be allergic, adequate labeling is essential.

10. The way forward

The recent surge in interest in insects as food and feed was sparked in part by the release of an FAO report in 2013, which has been downloaded more than 7 million times. Wageningen University and the FAO jointly organized the first conference on this topic, "Insects to Feed the World," in the Netherlands in 2014. This meeting drew 450 people from 45 different nations. In the agriculture, food, feed, and health sectors, research institutes, universities, private firms, international organizations, civil society, and government agencies were all represented [71].

Edible insects harvested from natural resources support livelihoods since they may be consumed and/or sold. Research on sustainable harvesting, semi-domestication, and farming is required to avoid overexploitation. Some insect pests, such as edible grasshoppers in Mexico, can be harvested as a management tool [72].

Insect farming veterinary science is still in its infancy. Insect diseases that may emerge during large-scale rearing are poorly understood in terms of biological and genetic characterization, phylogeny, host range, transmission, persistence, epidemic potential, and animal safety, including human safety [73]. Disease transmission has been a problem in the conventional livestock industry on a global scale. Microbial contamination prevention, detection, identification, and mitigation are critical for a successful and safe insect production. Insects can be used to convert organic waste streams like manure into high-protein goods, which is an intriguing prospect. However, further empirical investigations and monitoring are needed to determine the quality of the utilized garbage and the insects created. If insect-based food or feed is contaminated with dangerous bacteria, mycotoxins, or heavy metals, the potential risk to human health must be addressed immediately [73].

In the western world, legislative barriers are currently impeding the advancement of the emerging sector of insects as food and feed. The unfamiliarity with insects as food in the European Union, according to De-Magistris *et al.* [74], may impact EU decision-making since consumers are "conditioned" by cultural patterns and neophobia when it comes to edible insects. As a result, there may be less receptivity to new ideas. On the one hand, the EU investigates and promotes innovative and sustainable food ingredients such as insects, but on the other, it stifles innovation by imposing a regulatory framework that protects consumers from hazards associated with novel food items.

Because it is labor demanding and feed prices are considerable, concerns have been raised about the viability of mass-producing insects. To create vast numbers of high-quality and safe insect products in a cost-effective and reliable manner, automation of manufacturing operations would be required. Another option to save feedstock costs is to employ low-value organic by-products and waste streams [75].

A lot of firms are working on this project throughout the world. One firm can process 20 tons of fly larvae every day, yielding seven tons of insect meal and three tons of insect oil. Because vast amounts of feed are necessary for pets, fish, and cattle, and because the components for fishmeal and soymeal continue to rise in price, the use of insect meal as an alternative protein source is becoming a more attractive choice. More attention is needed to optimize desired insect features by selecting specific strains or utilizing genetic enhancement procedures [75].

Cultural and individual expectations regarding the species to be used as food and how they should be prepared should be considered when developing insect-based food products. It is inadequate to emphasize the health and environmental benefits to encourage usage. Gastronomy study is also required to determine whether insects are acceptable as a sustainable food source (deliciousness). Multiple disciplinary approaches (multi-disciplinarily, inter-disciplinarily, and trans-disciplinarily) are required to advance the new agricultural sector of insects as food and feed, as complex problems must be solved that transcend traditional boundaries and require the collaboration of non-academic stakeholders [75].

11. Conclusions

The biggest phylum in the animal kingdom, containing well-known insects, spiders, ticks, and crustaceans, as well as several smaller, lesser-known species and a plethora of bizarre forms only known as fossils. Arthropods make up over 95% of all animal species [1]. There are about one million recognized species, the majority of which are insects [1]. Nobody knows how many arthropod species there are. Some officials believe it may be as high as ten million [1]. The body of an adult arthropod is normally made up of a succession of ring-like segments with a pair of numerous jointed limbs on each segment that move on each other via muscles. It is becoming obvious that arthropod resources may be mass manufactured for use in food production for sustainable development. In terms of nutritional value, food components, and chemical makeup, it is a valuable resource. Meanwhile, the use of edible arthropods has posed a problem in terms of food security, environmental conservation, and the destruction of traditional culinary culture [1].

Acknowledgements

We wish to acknowledge the publishers, for the opportunity to add to the pool of knowledge, some drops of reviewed ideas. We hereby register our profound gratitude to the editors involved in publishing this work for the required quality and desired acceptability.

Author details

Jonathan Ibrahim[1*] and Dalyop Daniel Gyang[2]

1 Department of General Study, Gombe State College of Health Sciences and Technology, Kaltungo, Nigeria

2 Department of Pharmacy Technicians, Gombe State College of Health Sciences and Technology, Kaltungo, Nigeria

*Address all correspondence to: jibrahim@chstkaltungo.edu.ng

References

[1] Frederick RS, Jerome CR, Patsy AM. Arthropoda. USA: Access Science from McGraw-Hill Education; 2014

[2] Ojiako OA, Igwe CU. The nutritive, anti-nutritive and hepatotoxic properties of *Trichosanthes anguina* (Snake Tomato) fruits from Nigeria. Pakistan Journal of Nutrition. 2008;7:85-89

[3] Ojiako OA, Igwe CU, Agha NC, Ogbuji CA, Onwuliri VA. Protein and Amino acid compositions of *Sphenostylis stenocarpa, Sesamum indicum,Monodora myristica* and *Afzelia africana* seeds from Nigeria. Pakistan Journal of Nutrition. 2010;**9**:368-372

[4] Van Itterbeeck J, van Huis A. Environmental manipulation for edible insect procurement: a historical perspective. Journal of Ethnobiology and Ethnomedicine. 2012;**8**(3):1-19

[5] Jongema Y. List of Edible Insect Species of the World. Wageningen: Wageningen University; 2014

[6] Arnold VH. Edible insects contributing to food security? Journal of Agriculture and Food Security. 2015. DOI: 10.1186/s40066-015-0041-5. ISBN 978-925-107595-1

[7] Van Huis A, Van Itterbeeck KH, Mertens E, Halloran A, Muir G, et al. Edible Insects: Future Prospects for Food and Feed Security. Rome: Food and Agriculture Organization of the United Nations; 2013

[8] Makkar HPS, Tran G, Heuzé V, Ankers P. State-of-the-art on use of insects as animal feed. Animal Feed Science and Technology. 2014;**197**:1-33

[9] Kenis M, Koné N, Chrysostome CAAM, Devic E, Koko GKD, Clottey VA, et al. Insects used for animal feed in West Africa. Entomologia. 2014;**2**(218):107-114

[10] Arnold VH, Joost VI, Harmke K, Esther M, Afton H, Giulia M, et al. Edible insects Future prospects for food and feed security. FOA Forestry Paper 171. 2013

[11] Olsen RL, Hasan MR. A limited supply of fishmeal: impact on future increases in global aquaculture production. Trends in Food Science and Technology. 2012;**27**(2):120-128

[12] FAO. Amino Acid Content of Food and Biological Data on Proteins. Rome: Food and Agricultural Organization/ United National Joint Committee; 2014

[13] Sánchez-Muros MJ, Barroso FG, Manzano-Agugliaro F. Insect meal as renewable source of food for animal feeding: a review. Journal of Cleaner Production. 2014;**65**:16-27

[14] Lock ER, Arsiwalla T, Waagbø R. Insect larvae meal as an alternative source of nutrients in the diet of Atlantic salmon (*Salmo salar*) postsmolt. Aquatic Nutrition. 2015. DOI: 10.1111/anu.12343

[15] Rumpold BA, Schlüter OK. Nutritional composition and safety aspects of edible insects. Molecular Nutrient Food Research. 2013;**57**(5):802-823

[16] Rumpold BA, Schlüter OK. Potential and challenges of insects as an innovative source for food and feed production. Innovative Food Science Emergency Technology. 2013;**17**:1-11

[17] McLean E, Cogswell M, Egli I, Wojdyla D, De Benoist B. Worldwide prevalence of anaemia, WHO vitamin and mineral nutrition information

system. Public Health Nutrition. 2009;**12**(4):444-454

[18] Gibson RS. Dietary-induced zinc deficiency in low income countries: challenges and solutions the avanelle kirksey lecture at Purdue university. Nutrition Today. 2015;**50**(1):49-55

[19] Yi L, Lakemond CMM, Sagis LMC, Eisner-Schadler V, Van Huis A, Van Boekel MAJS. Extraction and characterization of protein fractions from five insect species. Food Chemistry. 2013;**141**(4):3341-3348

[20] Deng Y, Chen XM, Wang SY, et al. Common edible insects and nutritional value from food: How regulations and information can hamper radical innovations in the European Union. British Food Journal. 2010;**117**(6):1777-1792

[21] Rong BX, Gan SY. Ants and preparation of trace element analysis. Chinese Traditional Invertebrates Zoo Biology. 1998;**17**:123-134

[22] Van Huis A. Potential of insects as food and feed in assuring food security. Annual Review of Entomology. 2013;**58**(1):563-583

[23] DeFoliart G. Insect as human food; Gene DeFoliart discusses some nutritional and economic aspects. Journal of Crop Protection. 1992;**11**:395-399

[24] Verbeke W. Profiling consumers who are ready to adopt insects as a meat substitute in a Western society. Food Quality and Preference. 2015;**39**:147-155

[25] Charlton AJ, Dickinson M, Wakefield ME, Fitches E, Kenis M, Han R, et al. Exploring the chemical safety of fly larvae as a source of protein for animal feed. Journal of Insects Food and Feed. 2015;**1**(1):7-16

[26] Oliveira JS, de Carvalho JP, de Sousa RF, et al. The nutritional value of four species of insets consumed in Angola. Ecology of Food and Nutrition. 1976;**5**:91-97

[27] Rong BX, Gan SY. Ants and preparation of trace element analysis. Chinese Traditional and Herbal Drugs. 1987;**18**:47-49

[28] Zhou ZH, Yang W, Xu DP, et al. Analysis of Resource Value of Edible Insect. China: Science and Technology Press; 2002. pp. 71-146

[29] Wen LZ. Nutrition of edible insects in Mexico. Entomological Knowledge. 1998;**35**(1):58-61

[30] Liu XG, Ju XR, Wang HF, et al. Insect oil and its nourishment appraisement. Journal of the Chinese Cereals and Oils Association. 2003;**6**:11-13

[31] Matthew G, Schroeder FC. Insect natural products; comprehensive natural products II. Technology of Food Industry. 2014;**17**:390-394

[32] Klunder HC, Wolkers-Rooijackers J, Korpela JM, Nout MJR. Microbiological aspects of processing and storage of edible insects. Food Control. 2012;**26**:628-631

[33] Payne CLR. Wild harvesting declines as pesticides and imports rise: The collection and consumption of insects in contemporary rural Japan. Journal of Insects Food and Feed. 2015;**1**(1):57-65

[34] Payne CLR, Evans JD. Nested houses: Domestication dynamics of human-wasp relations in contemporary rural Japan. Journal of Ethnobiology and Ethnomedicine. 2017;**13**:13

[35] Akpalu W, Muchapondwa E, Zikhali P. Can the restrictive harvest

period policy conserve mopane worms in southern Africa? A bioeconomic modeling approach. Environmental Development Economy. 2009;**14**(5):587-600

[36] Hanboonsong Y, Jamjanya TT, Durst PB. Six-legged Livestock: Edible Insect Farming, Collection and Marketing in Thailand. Bangkok: Regional Office for Asia and the Pacific of the Food and Agriculture Organization of the United Nations; 2014

[37] Hanboonsong Y, Durst PB. Edible Insects in Lao PDR: Building on Tradition to Enhance Food Security. Bangkok: Regional Office for Asia and the Pacific of the Food and Agriculture Organization of the United Nations; 2014

[38] Ghazoul J. Mopani Woodlands and the Mopane Worm: Enhancing Rural Livelihoods and Resource Sustainability. London: DFID; 2006

[39] Alltech. Global Feed Survey. 2015. Available from: http://www.alltech.com/sites/

[40] Szelei J, Woodring J, Goettel MS, Duke G, Jousset FX, Liu KY, et al. Susceptibility of North-American and European crickets to *Acheta domesticus* densovirus (AdDNV) and associated epizootics. Journal of Invertebrate Pathology. 2011;**106**(3):394-399

[41] Gahukar RT. Edible insects farming: Efficiency and impact on family livelihood, food security, and environment compared with livestock and crops. 2016. pp. 85-111. DOI: 10.1016/B978-0-12-802856-8.00004-1

[42] Gahukar RT. Edible insects collected from forests for family livelihood and wellness of rural communities: A review. Global Food Security. 2020;**2020**:25

[43] Alamu OT, Amao AO, Nwokedi CI, Oke OA, Lawa IO. Diversity and nutritional status of edible insects in Nigeria: A review. International Journal of Biodiversity and Conservation. 2013;**5**(4):215-222

[44] Muafor FJ, Gnetegha AA, Philippe LG, Patrice L. Exploitation, Trade and Farming of Palm Weevil Grubs in Cameroon. Bogor, Indonesia: CIFOR; 2015

[45] Ebenebe CI, Okpoko VO. Preliminary studies on alternative substrate for multiplication of African palm weevil under captive management. Journal of Insects as Food and Feed. 2016;**2**(3):171-177

[46] Ande AT, Fasoranti JO. Some aspects of the biology, foraging and defensive behaviour of the emperor moth caterpillar, Cirina forda (Westwood). International Journal of Tropical Insect Science. 1998;**18**(03):177-181

[47] Van Huis A. Edible insects: Marketing the impossible? Journal of Insects Food Feed. 2017;**3**:67-68

[48] Netshifhethe SR, Kunjeku EC, Duncan FD. Human uses and indigenous knowledge of edible termites in Chombe district, Limpopo Province, South Africa. South African Journal Sciences. 2018;**2018**:11

[49] Mmari MW, Kinyuru JN, Laswai OK, Okoth JK. Traditions, beliefs and Indigenous technologies in connection with the edible longhorn grasshopper, Ruspolia differens (Serville) in Tanzania. Journal of Ethnobiology and Ethnomedicine. 2017;**13**:60

[50] Fombong FT, Van der Borght M, Broeck JV. Influence of freeze-drying and oven-drying post blanching on the nutrient composition of the edible insect, Ruspolia differens. Insects. 2017;**8**:102

[51] Schäufele I, Albores EB, Hamm U. The role of species for the acceptance

of edible insects: Evidence from a consumer survey. British Food Journal. 2019;**121**:2190-2204

[52] Siozios S, Massa A, Parr C, Verspar RL, Hurst GD. DNA barcoding reveals incoorect labelling of insects sold as food in the UK. Peer Journal. 2020;**8**:e8496. DOI: 10.7717/peerj.8496

[53] Meyer-Rochow VB, Hakko H. Can edible grasshoppers and silkworm pupae be tasted by humans when prevented to see and smell these insects? *Journal of Asian Paific*. Entomology. 2018;**21**:616-619

[54] Van Thielen L, Vermuyten S, Storms B, Rumpold B, van Campenhout L. Consumer acceptance of foods containing edible insects in Belgium two years after their introduction to the market. Journal Insects Food and Feed. 2019;**5**:35-44

[55] Barsics F, Megido RC, Brostaux Y, Barsics C, Blecker C, Haubruge E, et al. Could new information influence attitude to food supplemented with edible insects? British Food Journal. 2017;**119**:2027-2039

[56] Kewuyemi Y, Kesa H, Chinna CE, Adebo OA. Fermented edible insects for promoting food security in African. Insects. 2020;**11**:283

[57] Melghar-Lalanne G, Hernandez-Alvarez AJ, Salinas-Castro A. Edible insects processing: Traditional and innovative technologies. Comprehensive Review on Food Sciences and Food Safety. 2019;**2019**:18

[58] Son YJ, Choi SY, Hwang IK, Nho CW, Kim SH. Could defatted mealworm (Tenebrio molitor) and mealworm oil be used as food ingredients? Food. 2020;**9**:40

[59] Anuduang A, Loo YY, Jomduang S, Lim SJ, Mustapha WAW. Effect of thermal processing on physico-chemical and antioxidant propertie in mulberry silkworm (*Bombyx mori* L.) powder. Foods. 2020;**9**:871

[60] Williams JP, Williams JR, Kirabo A, Chester D, Peterson M. Nutrient content and health benefits of insects. In: Dossey AT, Morales-Ramos JA, Rojas MG, editors. Insects as Sustainable Food Ingredients: Production, Processing and Food Applications. New York, NY, USA: Elsevier Inc; 2016. pp. 61-84

[61] Chakravorty J, Ghosh S, Meyer-Rochow VB. Practices of entomophagy and entomotherapy by members of the Nyishi and Galo tribes, two ethnic groups of the state of Arunachal Pradesh (North East India). Journal of Ethnobiology and Ethnomedicine. 2011;**7**:5

[62] Meyer-Rochow VB, Chakravorty J. Notes on entomophagy and entomotherapy generally and information on the situation in India in particular. Applied Entomology and Zoology. 2013;**48**:105-112

[63] Gahukar RT. Edible insects collected from forests for family livelihood and wellness of rural communities: *A review*. Global Food Security. 2020;**25**:100348

[64] Nyangena DN, Christopher M, Samuel I, John K, Hippolyte A, Sunday E, et al. Effects of traditional processing techniques on the nutritional and microbiological quality of four edible insect species used for food and feed in East Africa. Journal of Foods. 2020;**9**(5):574

[65] Sabolová M, Adámková A, Kouˇrimská L, Chrpová D, Pánek J. Minor lipophilic compounds in edible insects. Potravinarsmt. 2016;**10**:400-406

[66] Bequaert J. Insects as food: How they have augmented the food supply of mankind in early and recent years. Natural Historical Journal. 2015;**21**:191-200

[67] Yen AL. Edible insects: traditional knowledge or western phobia? (Special Issue: Trends on the edible insects in Korea and abroad.). Entomological Research. 2009;**39**(5):289-298

[68] Kinyuru JN, Konyole SO, Owuor BO, Kenji GM, Onyango CA, Estambale BB, et al. Nutrient composition of four selected winged termites in western Kenya. Journal of Food Composition and Analysis. 2013;**30**(2):120-124

[69] Looy H, Dunkel FV, Wood JR. How then shall we eat? Insect-eating attitudes and sustainable foodways. Agricultural Human Values. 2014;**31**:131-141

[70] Tan HSG, Fischer ARH, Tinchan P, Stieger M, Steenbekkers LPA, Van Trijp HCM. Insects as food: exploring cultural exposure and individual experience as determinants of acceptance. Food Quality and Preference. 2015;**42**:78-89

[71] Verhoeckx KCM, Van Broekhoven S, Den Hartog-Jager CF, Gaspari M, DeJong GAH, Wichers HJ, et al. House dust mite (Der p 10) and crustacean allergic patients may react to food containing Yellow mealworm proteins. Food and Chemical Toxicology. 2014;**65**:364-373

[72] Pennisi E. All in the (bigger) family revised arthropod tree marries crustacean and insect fields. Science Total Environment. 2015;**347**(6219):220-221

[73] Van Huis A, Vantomme P. Conference report: insects to feed the world. Food Chain. 2014;**4**(2):184-192

[74] De-Magistris T, Pascucci S, Mitsopoulos D. Paying to see a bug on my food: How regulations and information can hamper radical innovations in the European Union. Britain Food Journal 2015;**117**(6):1777-1792

[75] Cerritos R, Cano-Santana Z. Harvesting grasshoppers *Sphenarium purpurascens* in Mexico for human consumption: a comparison with insecticidal control for managing pest outbreaks. Journal of Journal of Crop Protectionection. 2008;**27**(3-5):473-480

Chapter 10

Capybara Ticks and the Urban Context of Spotted Fever in Brazil: An Overview

Simone Magela Moreira, Ariane Flávia do Nascimento and Bruna Macena Pereira de Souza

Abstract

Spotted fever is caused by *Rickettsia rickettsii* and is transmitted through tick's saliva. Humans, ticks, and capybaras (*Hydrochoerus hydrochaeris*) are often coexisting in environments that favor the spread of Brazilian spotted fever (BSF). Although capybaras do not transmit *R. rickettsii*, they can amplify these bacteria among tick vector populations, playing a significant role in the one health approach and epidemiology of the disease. Urban populations of capybaras have increased, especially in Southeast Brazil, as well as the number of cases and lethality of BSF have increased in the country since the 1980s. This expansion is mainly determined by the availability of food and the absence of predators. Thus, urban areas, including parks and university campuses, provide an abundance of food and protection against predators, ensuring the multiplication of the species and increasing the risk of transmission to humans due to the proximity of man with animals in the urban environment. Therefore, this chapter aims to address aspects of spotted fever, considering the many dimensions of the species involved, contributing to public strategies and policies.

Keywords: urban disease, arthropod vectors, family Caviidae, rickettsiosis

1. Introduction

Brazilian Spotted Fever (BSF) also known as New World Spotted Fever or São Paulo Exanthematic Typhus [1] is becoming increasingly widespread among regions of Brazil. Since 2001, Brazil has reported the urbanization of BMF, particularly in the states of São Paulo and Minas Gerais where the disease is already considered endemic in many areas [2].

BSF is an infectious, multisystemic, and febrile disease caused by the species *Rickettsia rickettsii* and *R. parkeri*, which are the first responsible for the most serious manifestations, whose lethality can reach 80%, in situations of late diagnosis or lack of access to health services, common in developing countries. Although the expansion of occurrence spaces can have many causes, a commonly associated ecological element is the presence of capybaras, currently found in water bodies and lawns in urban areas [3]. They participate in the transmission cycle, being considered important amplifiers of vector ticks [4], affecting the endemicity and zoonotic aspects [5] of the occurrences. In fact, these rodents are spreading far beyond rural areas, forming clusters in leisure spaces, parks, and lakes present in highly urbanized cities, making infestation by ticks common in people who visit these places [6, 7].

Vector-borne diseases are relevant to human, animal, and environmental health, as pathogens, vectors, and hosts interact through pathologies and can change their epidemiology over time [8]. When considering the changes promoted by the exploitation of natural resources that result in the fragmentation of habitats and changes in ecosystems, it is possible to assume that there is a greater interaction between humans and arthropod-borne pathogens.

In this context, the global strategic framework for health, created to mitigate the risk and minimize the impact of emerging infectious diseases at the animal-human-ecosystem interface and socio-economy, highlights the need for action in One Health Perspective that fits perfectly with the approaches relating to the BSF. Based on the principles of "one world," "one health," the referred framework resulted from an action between specialized agencies, such as the World Health Organization (WHO), the United Nations (UN) Food and Agriculture Organization (FAO), and the International Organization for Animal Health (OIE), which jointly recognized the link between human welfare, animals, and the environment [9]. Diseases involving humans and animals (domestic and wild) should be addressed as a priority in an interdisciplinary manner to combat threats and promote the health of life on Earth [10]. The approach must be holistic for disease prevention and for the integrity of the ecosystem that sustains all forms of life.

Due to the complexity of the parasitic cycle involving different hosts (wild animals, arthropod vectors, and humans), the conditions and variables that affect interactions are not completely understood, particularly when related to vector diseases. Therefore, the present narrative intends to provide some clues for a better understanding of some of the main characteristics associated with the occurrence of BSF.

2. Vector-borne diseases in the urban environment

In 1950, only 30% of the world's population lived in urban areas. World's population is increasingly urban with more than half living in urban areas. This number continues to increase, and by 2050, two-thirds are expected to be living in newly urbanized areas [11]. However, the size and density of human populations are creating challenges for many dimensions of human health, including the (re)emergence of zoonotic diseases, particularly those transmitted by vectors [12].

Changes in landscape structure stimulate the loss, rotation, or homogenization of biodiversity, increasing the contact between humans and animals, which can influence the dynamics of transmissions in communities of reservoir hosts and vectors, contributing to modulations in epidemiological processes and the expansion of pathogens that reach the humans [13, 14]. The beginning of the 2020s was marked by a serious pandemic that initially presented itself with zoonotic characteristics, whose cause corroborates the new thinking that as humans spread, pathogens also spread, representing an even greater problem in countries with a large number of poor people and in emerging economies. This is because, in such locations, in addition to much nature being transformed into agricultural or urban areas, health systems are usually underfunded and have difficulties in dealing with possible outbreaks [15].

Cities, as densely populated areas, have always been habitats for domestic animals. For some time, however, they are increasingly housing "wild" animals as they present themselves as safe and well-supplied places where these animals have chosen to survive and raise their descendants. This is because, in their decisions, when choosing a residence, food and security are more attractive than other aspects of the landscape [16]. As a result, a park close to busy streets, with noisy children, has been more attractive to these animals than the curious quietness of rural landscapes, "drenched in chemicals" [17].

Of course, given the complexities of ecosystems and rapid global changes, the effects of land use on the ecological factors that sustain zoonotic occurrences cannot be hastily judged [18]. Although there is evidence that the diversity of local species affects the transmission of pathogens [19], this result is not generic, with the greatest risks for diseases caused by pathogens transmitted by arthropod vectors being proven [20, 21], as in the BSF cycle.

The reason why such pathogens spread after environmental changes is not that we suddenly come into contact with wild animals, such as jaguars and wolves. This hypothesis does not answer the modern questions that involve the maintenance, for decades, of these zoonotic cycles in the most anthropized environments. On the contrary, researchers reveal that most common wild animals, particularly those known to carry pathogens threatening to humans, are more prolific in these areas [15]. However, the biases that cooperate for the intense multiplication of these vectors after the urbanization process are not yet fully clarified. Surprisingly, animals that generally carry many viruses or bacteria seem to better tolerate the destruction of nature, when compared with those that carry fewer pathogens [22]. One reason for this could be that these animals are usually quite small and short-lived or have adaptive immune strategies that do not let them get sick from the pathogens they host.

The tendency for hosts and non-hosts to respond differently to human-induced changes in their habitats has been observed in some specific disease systems, but may contribute to the documented links between anthropogenic ecosystems and emerging zoonoses [21, 23] and for BSF, transmitted in an urban cycle that involves ticks, capybaras, and humans. Hosts with an accelerated life cycle contribute to an increase in their abundance, virtually due to life history trade-offs—between reproductive rate and investment— [24], associated with the ability to be resilient in the face of anthropic pressures [25]. In addition, characteristics such as host status, human tolerance [26], as well as the challenge load and co-evolution of shared pathogens [13] contribute to the clarification of such interactions.

3. Species involved in the transmission of spotted fever in Brazil

BSF is caused by Gram-negative bacteria of the genus Rickettsia (Rickettsiales: Rickettsiaceae), transmitted by vectors that use different vertebrate hosts, altering the endemicity and zoonotic aspects of their occurrence. Due to their strictly intracellular survival, these bacteria are classically transmitted to humans by arthropods, which include ticks, mites, fleas, and lice. However, many non-pathogenic human species have been described, of which the true roles in the ecological relationships with the vectors and with the pathogenic rickettsiae have not yet been fully clarified [27].

Since 2001, Brazil has reported an expansion of transmission areas, which is seriously integrated with the increase in the number of reported cases and in their lethality. The bioagent *R. rickettsii* is the main species causing BSF, being restricted to the Americas with confirmations, in addition to Brazil, Canada, United States, Mexico, Costa Rica, Panama, Colombia, and Argentina. However, in the country, other rickettsiosis from the spotted fever group has already been isolated, such as *Rickettsia parkeri* cepa Mata Atlântica and *R. parkeri stricto sensu*, whose notifications are associated with less severe and, as a rule, non-lethal clinical conditions [28, 29]. Added to this diversity of infectious agents, there is a variety of potential vectors that make the enzootic and epidemic cycle of BMF in Brazil quite complex, as a result of possible eco-epidemiological variations.

Ticks can act as vectors and reservoirs in the transmission dynamics of pathogens from the spotted fever group. Several species of Rickettsia can coexist in the same environment, so that different species of ticks that parasitize different mammals can also become infected. The main ticks implicated in the transmission of *R. rickettsii* in the United States are *Dermacentor andersoni* and *D. variablilis*; in some areas in Mexico and in the state of Arizona in the United States, *Rhipicephalus sanguineus* has been indicted as the vector; in South America, the tick species *Amblyomma cajennense* is the most commonly incriminated vector.

However, in the context of the BSF, the following interactions are currently described: [i] *R. rickettsii* transmitted by the tick *Amblyomma sculptum* and *Amblyomma aureolatum* in the Southeast and parts of the South region; [ii] *R. parkeri* Atlantic Forest strain vectored by *Amblyomma ovale* in Atlantic Forest fragments in the South, Southeast, and Northeast of the country; and [iii] *Amblyomma tigrinum* infected with R. *parkeri* is in the Pampa biome in the South region, distinguishing these Brazilian areas from others in Latin America [28, 30, 31].

Below, images of the two main tick species associated with the transmission of the bacterium *R. rickettsii*, which causes Brazilian spotted fever (**Figure 1**).

Bacterial infection in arthropods occurs during hematophagy performed on a rickettsemic, vertebrate host; being favored by transovarian and/or transstadial transmission, which balances part of the damage caused by infection in vector ticks [33]. In turn, the transfer of the pathogen to another vertebrate occurs when the infected ectoparasite takes a new blood meal, after the changes in phases. Thus, the occurrence of Rickettsia in a given space is based on the coexistence of ixodid species

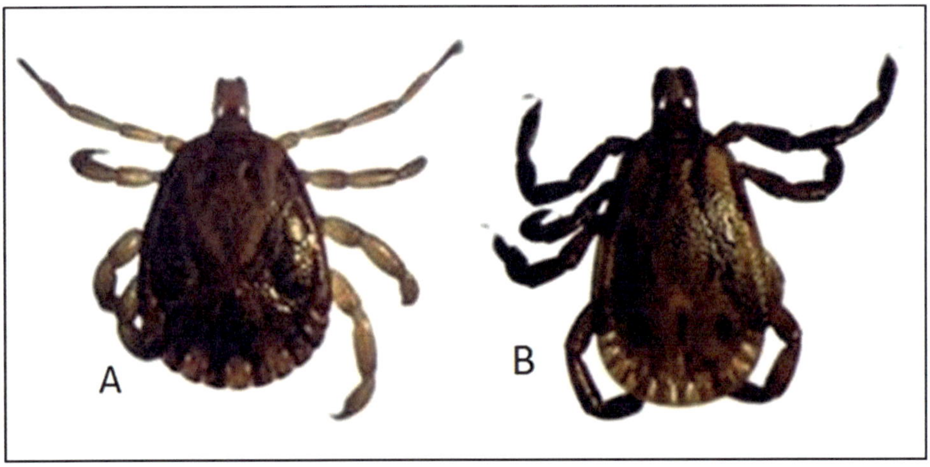

Figure 1.
Adult male ticks of the species (A) A. sculptum and (B) A. aureolatum. Source: [32].

susceptible to infection and on vertebrates capable of sustaining this tick's population. Both can vary over time and space, influencing the epidemic cycle by overlapping human activities with the proximity of other vertebrate hosts and the seasonality of the tick [30, 34].

In this aspect, some key factors in the epidemiology of BSF, particularly the occurrence of *R. rickettsii* in southeastern Brazil, are already partially clarified and point to the need for amplifying hosts to maintain transmission. Larvae, nymphs, and adults of *A. sculptum* are partially refractory to rickettsia and less than half of infected females are able to promote transovarian transfer (transmission to offspring) effectively. In addition, higher mortality and lower reproductive performance are observed when the tick is infected, compared with those free from the pathogen [35]. Given this finding, mathematical models indicated that *A. sculptum* cannot sustain *R. rickettsii* for successive generations without the genesis of new cohorts of infected ticks, via horizontal transmission, made possible by vertebrate amplifying hosts, during rickettsemia.

Horses are primary hosts of the *Amblyoma sculptum* tick. They act in the maintenance of these arthropods and in the movement of rickettsiae between environments. Even though they are not susceptible to infection, horses serve as sentinels in epidemiological studies, showing the distribution of the disease and predicting human cases [36]. However, in the contexts of the occurrences, the capybara (*Hydrochoerus hydrochaeris*) appears as the largest amplifying host of *R. rickettsii* for *A. sculptum* in BSF endemic areas in southeastern Brazil [7, 37–39]. In the state of São Paulo, some areas became endemic for BSF after the detection of an increase in the number of individuals in free-ranging capybara clusters [40].

Thus, the sanitary monitoring of capybara populations is essential for the control and conservation of public health. For the record, it is worth mentioning that the tick *Amblyomma dubitatum* has also been frequently discovered infesting capybaras in southeastern Brazil, despite not playing an essential role in the epidemiology of BSF [41–43].

4. Aspects of the epidemiological scenario

The first report of spotted fever in Brazil occurred in São Paulo, in 1929 [1], and *R. rickettsii* transmitted by the tick A. cajennense was indicated as the causal agent. In the report, the similarities with Rocky Mountain spotted fever were already highlighted. In the following two decades, new occurrences were described in the state of Minas Gerais [44] and in São Paulo, where the species *Rickettsia typhi* was isolated for the first time [45]. After this period, BSF remained for several years as a silent disease, subordinated to existing flaws in medical care, diagnosis, and information processes.

Between the 1980s and 2000s, case reports are seen in scientific journals, indicating infections that occurred in southeastern Brazilian states, after four cases in Rio de Janeiro [46] and in the state of Espírito Santo, which came to be considered as an endemic region for BSF [47]. The states of Minas Gerais and São Paulo showed a serious reemergence, in whose epidemic outbreaks occurred in 1984, 1992, 1995, and the lethality reached about 50% of diagnosed patients [48–50]. Despite this, it was only in 2001 that the BSF notification became mandatory and, in 2014, it was determined that, in addition to being mandatory, it must be immediate, taking place within a maximum period of 24 hours, in order to improve surveillance, diagnosis, and treatment [51].

Due to the continuous increase in cases and the expansion of the areas of occurrence, BSF was considered an emerging disease in Brazil, without, however, a significant advance in information and knowledge about the disease, among the scientific, medical classes, and the general population [31]. Even today, there is not enough data to determine the impact of BSF on the Brazilian population, since prospective longitudinal studies documenting the natural course of the disease have not been performed.

In the brief history described above, two important aspects stand out in the epidemiological context. Failures in the diagnosis were confirmed through laboratory documents in the various fatal cases that occurred in Minas Gerais until the early 2000s [52]. Equally important, in 1996, Lemos et al. [53] isolated bacteria from the group of *Rickettsia* sp. of spotted fever in the tick *Amblyomma cooperi*, vectors collected from capybaras in São Paulo. It is clear that the impact of the BSF on public health will not be properly evaluated until appropriate methods of diagnosis and epidemiological surveillance are effectively implemented. Recent data bring some clarity about the incidence rates and some aspects (i) of humans affected by the disease; (ii) ticks (maintained by capybaras); and (iii) transmission environments, and need to be analyzed [54, 55]. The transversality between these aspects will be briefly described below.

In the Brazilian occurrences of BSF, white men, aged between 20 and 64 years, coming from rural areas are more commonly infected and report having had contact with ticks during their leisure activities, unlike women who have a domestic and peridomestic environment, as the likely site of infection [56]. Most cases affecting males corroborate the data described in international reports from the Centers for Disease Control and Prevention [57] and the European Center for Disease Prevention and Control [58]. The fact that most cases are represented by white ethnicity may be influenced by the difficulty in observing the rash when in black individuals, and further compromise the diagnosis of BSF in this ethnic group [59].

Studies in endemic areas of BSF have shown that the disease is correlated with environments where there are large populations of the tick *A. sculptum*, which, in turn, are supported by the presence of capybaras, fundamental hosts for this species, in anthropic environments. On the contrary, landscapes similar to these, where there are low parasite loads or where the tick *A. dubitatum* is predominant, did not become endemic areas, even with the presence of capybaras [7]. Recently, research conducted by Geraldi et al. [35] revealed that there are other aspects that contribute to the dynamics of transmission by demonstrating that there are variations in the susceptibility of the tick to infection by *R. rickettsii*, which may explain the different frequencies of BSF in areas where the vector A. sculptum and capybaras cohabit. However, the complete elucidation of the mechanisms that govern this susceptibility and their effects on the risk of the disease in urban environments still need further studies [7].

Historically, BSF notifications predominate in southeastern Brazil, with the percentage of fatal cases ranging from 30–50% [60]. Particularly in the state of São Paulo, there are extensive areas where transmission has been proven, with 978 laboratory confirmed cases between 2001 and 2018 [7]. The disease, however, continues to advance through new territories, reaching the south, northeast, and central-west regions of Brazil [56], varying estimates on the prevalence and incidence (**Figure 2**).

Some factors seem to influence the lethality percentages. Undeniably, the strain that prevails in Brazil is much more virulent when compared with the *R. rickettsii* that occurs in North America, responsible for Rocky Mountain spotted fever [61]. However,

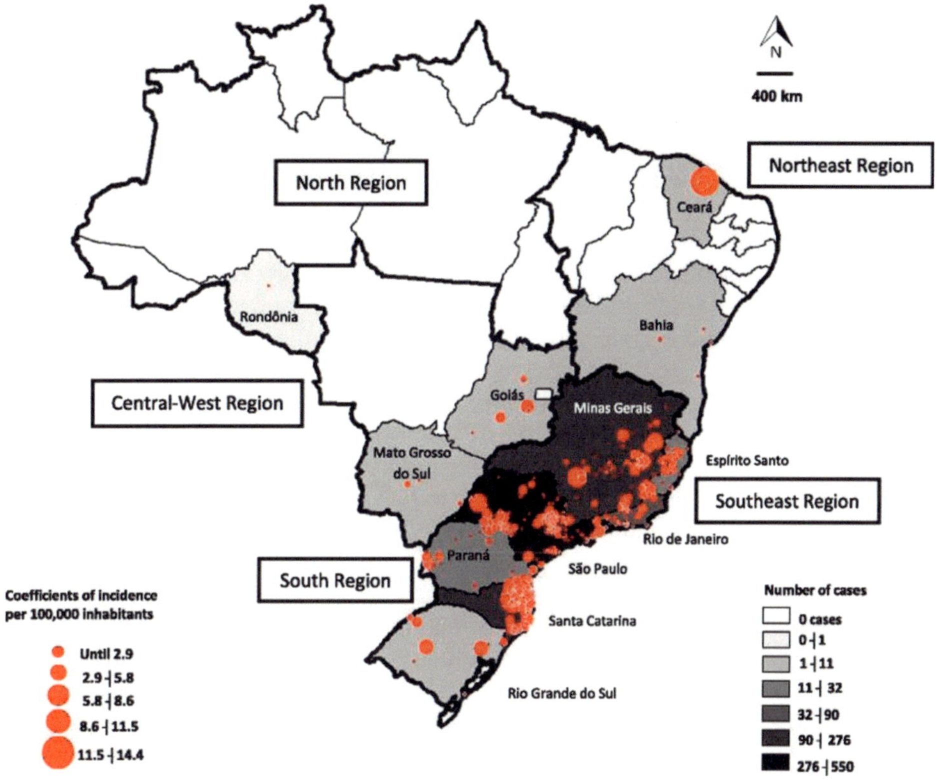

Figure 2.
Geographical distribution of confirmed cases of spotted fever by federative unit (FU) and average incidence rate in affected municipalities and between 2007. Source: [56].

aspects such as the lack of specificity in its clinical signs and the absence of exanthema (classic marker) in many patients lead to suspicion and make early diagnosis difficult. Diseases such as dengue, viral exanthematous diseases, and leptospirosis present clinical signs similar to those of BSF, requiring specificities in the differential diagnosis and increasing the possibility of the use of inappropriate drugs that prevent therapeutic cure [62]. Furthermore, it should not be forgotten that infections caused by R. parkeri, a strain present in several regions of the Atlantic Forest, present mild clinical signs, with inoculation bedsores and lymphadenopathy, introducing other elements that make adequate medical management difficult [63]. Furthermore, the geographically restricted availability of parenteral doxycycline, the drug of choice for the treatment of severe cases, highlights the neglected status of BSF in Brazil [64, 65]. A review of the treatment protocol for BSF took place in 2013 and doxycycline is now recommended regardless of patient age, being provided by the Ministry of Health for strategic locations in endemic areas. However, in silent areas or with recent introduction of the agent, the difficulty in accessing the drug remains [66].

In Brazil, rapid environmental changes, the absence of stable policies for environmental preservation, and the low socioeconomic condition of the population are associated with the high importance of BSF—expressed by the high lethality in outbreaks that predominate in family nuclei—establishing this Rickettsiosis as a relevant public health problem. Therefore, continued studies and solutions to support surveillance strategies will help in future assessments of epidemiological aspects.

5. The growing urbanization of capybaras and Brazilian spotted fever

Mammals represent the most successful evolutionary class among vertebrates. Facilitated by a brain that promotes learning and being capable to maintain a constant body temperature, they developed, throughout evolution, a variety of life strategies that allowed them to colonize the most diverse habitats, establishing themselves on all continents [67].

H. hydrochaeris is among the largest rodents found in the Neotropics. When Iberian settlers arrived in South America at the end of the fifteenth century, they came across a diverse fauna and the species was originally named, after analogies with other European animals known to them. However, the name capybara originates from the indigenous word (Tupi): kapii'gwara, which means grass eater (ka'pii = "grass" + g wara = "eater") [68]. It had its first detailed description in the mid-seveteenth century, based on observations in the state of Pernambuco, when northeastern Brazil was occupied by the Dutch. Later, in the search to define their origin and their preferred habitats, mistakes were made, but according to [69], the São Francisco River, one of the main Brazilian rivers, should be listed as the typical locality for the species, as mentioned by [70].

In Brazil, capybaras can be found in all 26 states and the Federal District (**Figure 3**) mainly in agricultural habitats, with a predominance of pastures and sugarcane fields where they can reach high densities. They are considered competitors that cause damage to a variety of crops, including sugarcane, corn, rice, bananas, soybeans, and compete for food with cattle, affecting agricultural production [70, 71]. In addition to these places, they frequent bodies of water (rivers, dams, and reservoirs) within urban limits, in public parks, and residential areas [69], currently causing conflicts, particularly in the Southeast, where they invade properties, eat ornamental plants in gardens, and are involved in traffic accidents on the streets and roads [3].

Recent increases in BSF cases [72] and the probable association with high capybara densities have accentuated considerations about the epidemiological role of this rodent in the urbanization of the disease. The expansion of occupation areas occurs as agricultural deforestation occurs, mainly due to the availability of food and the decline of their natural predators, such as jaguars [48, 49]. Thus, they can form numerous populations in certain environments, coming to be considered urban pests. This is because, in anthropogenic wetlands, where it finds an abundant source of food, *H. hydrochaeris* can develop with a carrying capacity greater than that observed in untouched environments. These facts raise important points about the roles played and the risk of transmission in important zoonosis, such as BSF [50].

It is also worth emphasizing relevant aspects regarding the presence of animals in urban daily life: the presence of wild animals in human groups can be interpreted as a possibility of greater exposure to nature and a source of benefits for the mental health of individuals [73], and increasing the value of recreational ecosystem services provided by green areas [74]. The relationship with nature offers benefits, even if there is no handling or prolonged intimacy in contact with animals. Simple eye contact, regardless of duration, can have a broad and robust impact on people's affective and cognitive conditions [73, 75].

According to the biophilia hypothesis [76], as a result of evolution, humans have always sought to connect with other life forms. This hypothesis translated some of the multiple dimensions of humans' innate relationship with nature, including emotional connections with landscapes and animals. In the last decade,

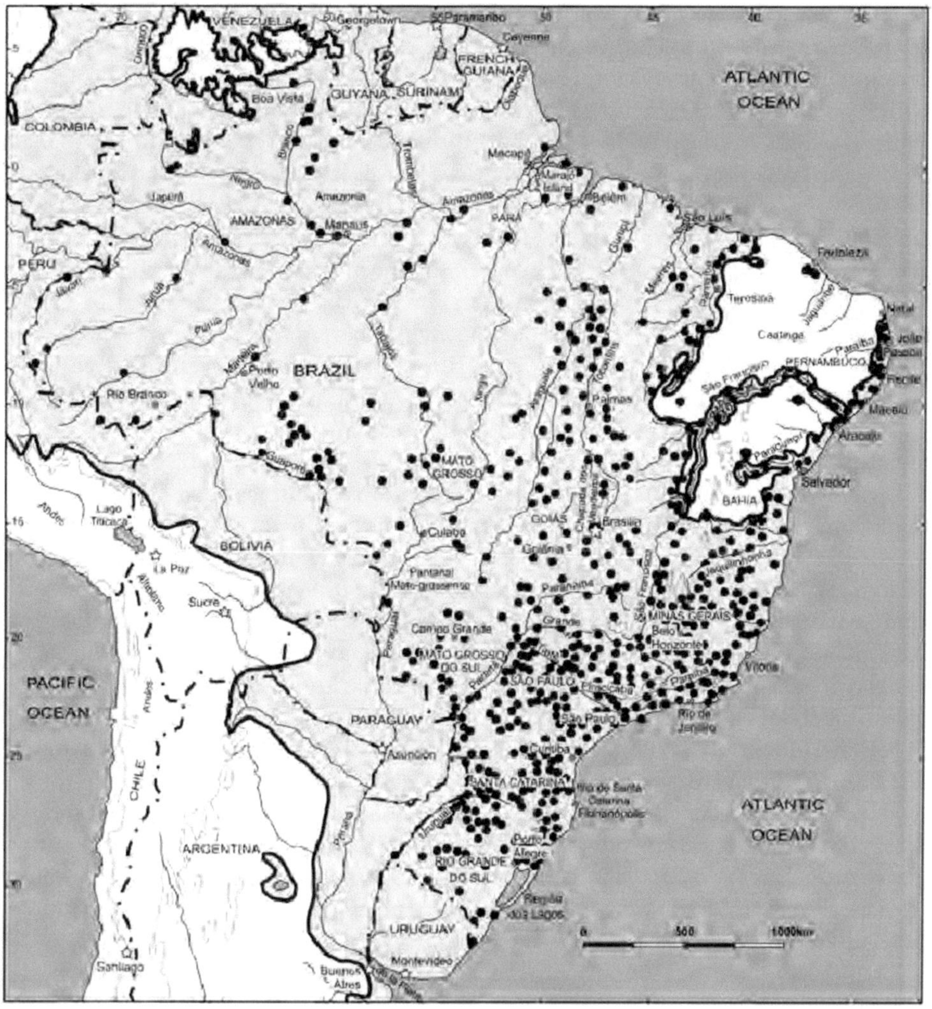

Figure 3.
Distribution of capybaras (Hydrochoerus hydrochaeris) in Brazil. Black dots show records of the species' presence. Source: [69].

the benefits of this contact have been increasingly studied [77]. Improvements in neuropsychological development and mental health have been reported when experiences occur in early childhood [78, 79], but benefits such as reductions in social and emotional difficulties and even deterioration cognitive impairment can extend into old age [80].

Urbanization, as it threatens to weaken the link between humans and nature [81], converging with Wilson's proposals relating to biophilia, may favor (or even promote) the acceptance of capybaras in the enclosures of a City. However, it is worth mentioning that as capybaras constitute groups that occupy leisure spaces, parks, and lakes, which, when located in areas with occurrence of BSF, increase the risk of tick infestations among people who visit these places [7, 82].

Although capybaras cannot transmit *R. rickettsii*, they play an important role in public health, as they amplify the agent among tick populations [4]. Thus, despite the presence of capybaras increasing biodiversity in cities and reinforcing the biophilia hypothesis, they pose a risk in the transmission of pathogens, requiring interdisciplinary participation and integrated actions to improve disease control. Researchers have already proven that BSF-endemic areas have much higher tick loads both in capybaras and in the environment, when compared with areas where the disease is not frequent [7]. And, to make the scenario even more complex, mathematical models have shown that the introduction of a single capybara infected with *R. rickettsii* parasitized by at least one infected tick is enough to establish an infection by *R. rickettsii* in the entire population of *A. sculptum* hosted by up to 50 capybaras [37]. In maintaining the infection, it is noteworthy that a capybara, during the primary infection, can remain in bacteremia for about 14 days, infecting other ticks that feed on its blood during this period [35].

The tick *A. sculptum* infected by *R. rickettsii* remains infected even after ecdysis (changes of stages), being able to transmit the bacteria in the following stages [35]. Thus, even with its preferred hosts in endemic areas, *A. sculptum* can accidentally parasitize humans in all its active stages [51], completing the zoonotic cycle.

Transmission to humans depends on the duration of contact between the tick and the person, requiring at least 4–6 hours for transmission to occur [5]. However, as very small tick stages can transmit rickettsiae, it is very common for people to remain infested long enough with the ticks without realizing it. Symptoms appear between 2 and 14 days after infection [83] and are mostly nonspecific. Early treatment is essential and mortality is associated with the difficulty in establishing the diagnosis, the delay in starting the specific therapy, and the little knowledge of the medical profession about the disease [84].

Tetracyclines and chloramphenicol are the only drugs with proven efficacy to treat BSF. In adults, treatment requires high dosage and needs to be continued for a few days after the fever subsides. Thus, in Brazil, despite the high prevalence and the fact that many areas are considered endemic for BSF, the lack of knowledge about the disease often leads to a delay in diagnosis, complicating the prognosis [85]. Therefore, it is important to take advantage of every opportunity to disseminate and expand knowledge about this disease.

Finally, it is worth noting that capybaras and ticks are only parts of the BSF cycle, requiring a deeper understanding of the different aspects of ecological relationships and monitoring of environmental conditions to reduce infections in humans [86]. We consider that cities are moldable and serve as a habitat for people and animals, and it is essential to analyze the extent to which urban ecology has been concerned with offering opportunities for new and different forms of interaction in human-animal relationships. According to the French anthropologist Philippe Descola [87], it is becoming increasingly clear that the established concepts—since the Renaissance—about city and countryside; culture and nature or humans and animals are not sustainable and need to be revisited in contemporary times.

6. Final considerations

In terms of future perspectives, given the persistence of the urbanization pattern, it seems clear that the BSF will not disappear easily from Brazil. In view of the above, there is a need for constant and rigorous epidemiological surveillance in urban areas where capybara is present. However, in the management of the conflicts mentioned above, disinformation must be fought [60] and short-term solutions are not inefficient. Public policies must adapt to manage the issue of BSF, in its urban occurrences, taking into account the many dimensions of the relationships between humans, other animal species, and the environment.

In order to reduce the risk of transmission and minimize the impact of BSF, surveillance and response systems should be instituted at national and regional levels, supporting public and animal health services with strategies to protect the health of ecosystems.

In addition, the greater lethality of BSF is also evidenced by its neglected status in Brazil, such as the lack of information in the health and surveillance sectors and the unavailability, in many regions of the country, of first-choice medication for the treatment of severe clinical cases of the disease.

We understand that the issue of the BSF and all aspects that involve its control must be rethought, correcting interventions not directed to health, with the development of feasible and accessible analysis methodologies within an integrative, multidisciplinary, and multisectoral vision with all that, directly or indirectly, have to do with this problem, in an articulation of planning, governance, and public health, in the search for health cities.

Finally, we emphasize the need to change the current emphasis on the short-term response to the disease and to encourage the construction of more sustainable systems capable of responding effectively to future events that involve the various nuances of the BSF that arise from the environment-animal-human interface.

Author details

Simone Magela Moreira, Ariane Flávia do Nascimento*
and Bruna Macena Pereira de Souza
Federal Institute of Minas Gerais, Bambuí, Minas Gerais, Brazil

*Address all correspondence to: ariane.nascimento@ifmg.edu.br

References

[1] Piza JT, Meyer JR, Gomes LS. Typho Exanthematico de São Paulo. São Paulo: Sociedade Impressora Paulista; 1932

[2] Souza MO. Multiscale analysis for a vector-borne epidemic model. Journal of Mathematical Biology. 2014;**68**(5):1269-1293. DOI: 10.1007/s00285-013-0666-6

[3] Bovo AA, KMPMB F, Verdade LM, Moreira JR. Capybaras (Hydrochoerus hydrochaeris) in Anthropogenic Environments: Challenges and Conflict2016. pp. 118-189. DOI: 10.1515/9783110480849-013

[4] Labruna MB. Brazilian spotted fever: The role of capybaras. In: Moreira JR, Ferraz KMPMB, Herrera EA, et al., edição. Capybara: Biology, Use and Conservation of an Exceptional Neotropical Species. 2013. pp. 371-383. DOI: 10.1590/1089-6891v18e-44671

[5] Darby AC, Cho NH, Fuxelius HH, Westberg J, Andersson SGE. Intracellular pathogens go extreme: Genome evolution in the Rickettsiales. Trends in Genetics. 2007;**27**(10):511-520. DOI: 10.1016/j.tig.2007.08.002

[6] Medeiros AP, Souza AP, de Moura AB, Lavina MS, Bellato V, Sartor AA, et al. Spotted fever group Rickettsia infecting ticks (Acari: Ixodidae) in the state of Santa Catarina, Brazil. Memórias do Instituto Oswaldo Cruz. 2011;**106**(8):926-930. DOI: 10.1590/S0074-02762011000800005

[7] Luz HR, Costa FB, Benatti HR, Ramos VN, Serpa MCA, Martins TF, et al. Epidemiology of capybara-associated Brazilian spotted fever. PLoS Neglected Tropical Diseases. 2019;**13**(9):e0007734. DOI: 10.1371/journal.pntd.0007734

[8] Harrus S, Baneth G. Drivers for the emergence and re-emergence of vector-borne protozoal and bacterial diseases. International Journal for Parasitology. 2005;**35**(11-12):1309-1318. DOI: 10.1016/j.ijpara.2005.06.005

[9] World Organisation for Animal Health (OIE) Contributing to One World, One Health. A Strategic Framework for Reducing Risks of Infectious Diseases at the Animal-Human-Ecosystems Interface. Paris, France: World Organization for Animal Health (OIE); 2008 https://www.fao.org/3/aj137e/aj137e00.pdf

[10] Wildlife Conservation Society. One World, One Health: Building Interdisciplinary Bridges to Health in a Globalized World Conference. 29th September 2004, The Rockefeller University, New York, NY. 2004. http://www.oneworldonehealth.org/sept2004/owoh_sept04.html

[11] United Nations, Department of Economic and Social Affairs, Population Division. World Urbanization Prospects: The 2014 Revision, Highlights (ST/ESA/SER.A/352)

[12] Myers SS et al. Human health impacts of ecosystem alteration. Proceedings of the National Academy of Sciences of the United States of America. 2013;**110**:18753-18760

[13] Plowright RK et al. Pathways to zoonotic spillover. Nature Reviews Microbiology. 2017;**15**:502-510. DOI: 10.1038/nrmicro.2017.45

[14] Shah HA, Huxley P, Elmes J, Murray KA. Agricultural land-uses consistently exacerbate infectious disease risks in Southeast Asia. Nature

Communications. 2019;**10**:4299. DOI: 10.1038/s41467-019-12333-z

[15] Gibb R, Redding DW, Chin KQ, et al. Zoonotic host diversity increases in human-dominated ecosystems. Nature. 2020;**584**:398-402

[16] Morizot B. Philosophie der Wildnis oder die Kunst vom Weg abzukommen. Reclam Philipp Jun. Ditzingen2020. p. 191

[17] Volker A, Claus-Peter H. Das Verstummen der Natur. Das unheimliche Verschwinden der Insekten, Vögel und Pflanzen – und wie wir es noch aufhalten können. Bonn. 2019. p. 336.

[18] Gottdenker NL, Streicker DG, Faust CL, Carroll CR. Anthropogenic land use change and infectious diseases: A review of the evidence. EcoHealth. 2014;**11**:619-632

[19] Civitello DJ et al. Biodiversity inhibits parasites: Broad evidence for the dilution effect. Proceedings of the National Academy of Sciences of the United States of America. 2015;**112**:8667-8671

[20] LoGiudice K, Ostfeld RS, Schmidt KA, Keesing F. The ecology of infectious disease: Effects of host diversity and community composition on Lyme disease risk. Proceedings of the National Academy of Sciences of the United States of America. 2003;**100**:567-571

[21] Kilpatrick AM. Globalization, land use, and the invasion of West Nile virus. Science. 2011;**334**:323-327

[22] Kamiya T, O'Dwyer K, Nakagawa S, Poulin R. What determines species richness of parasitic organisms? A meta-analysis across animal, plant and fungal hosts. Biological reviews of the Cambridge Philosophical Society. 2014;**89**:123-134

[23] Pulliam JRC et al. Agricultural intensification, priming for persistence and the emergence of Nipah virus: A lethal bat-borne zoonosis. Journal of The Royal Society Interface. 2012;**9**:89-101

[24] Lee KA, Wikelski M, Robinson WD, Robinson TR, Klasing KC. Constitutive immune defences correlate with life-history variables in tropical birds. Journal of Animal Ecology. 2008;**77**:356-363

[25] Purvis A, Gittleman JL, Cowlishaw G, Mace GM. Predicting extinction risk in declining species. Proceedings of the Biological Society. 2000;**267**:1947-1952

[26] Joseph MB, Mihaljevic JR, Orlofske SA, Paull SH. Does life history mediate changing disease risk when communities disassemble? Ecology Letters. 2013;**16**:1405-1412

[27] Burgdorfer W. Considerações ecológicas e epidemiológicas da febre maculosa das Montanhas Rochosas e do tifo do matagal. Em: DH Walker edição. Biologia de doenças rickettsiais. Vol. 1. 1988. pp. 33-50

[28] Parola P, Paddock CD, Socolovschi C, Labruna MB, Mediannikov O, Kernif T, et al. Update on tick-borne rickettsioses around the world: A geographic approach. Clinical Microbiology Reviews. 2013:657-702. DOI: 10.1128/CMR.00032-13

[29] Weck B, Dall'Agnol B, Souza U, Webster A, Stenzel B, Klafke G, et al. Spotted fever group Rickettsia in the Pampa biome, Brazil, 2015-2016. Emerging Infectious Diseases. 2016;**22**:2014-2016. DOI: 10.3201/eid2211.160859

[30] Szabó MPJ, Pinter A, Labruna MB. Ecology, biology and distribution of spotted-fever tick vectors in Brazil.

Frontiers in Cellular and Infection Microbiology. 2013;**3**:1-9. DOI: 10.3389/fcimb.2013.00027

[31] Oliveira SV, Guimarães JN, Reckziegel GC, da BMC N, de KM A-V, Fonseca LX, et al. An update on the epidemiological situation of spotted fever in Brazil. Journal of Venomous Animals and Toxins including Tropical Diseases. 2016;7:65-72. DOI: 10.1186/s40409-016-0077-4

[32] Moraes-Filho J. Brazilian spotted fever. Journal of Continuing Education in Animal Science of CRMV-SP. 2017;**15**(1):38-45

[33] Eremeeva ME, Dasch GA. Challenges posed by tick-borne rickettsiae: Eco-epidemiology and public health implications. Frontiers in Public Health. 2015;**3**:55. DOI: 10.3389/fpubh.2015.00055

[34] Szabó MPJ, Nieri-Bastos FA, Spolidorio MG, Martins TF, Barbieri AM, Labruna M. In vitro isolation from Amblyomma ovale (Acari: Ixodidae) and ecological aspects of the Atlantic Rainforest Rickettsia, the causative agent of a novel spotted fever rickettsiosis in Brazil. Parasitology. 2013;**140**:719-728. DOI: 10.1017/S0031182012002065

[35] Gerardi M, Ramírez-Hernández A, Binder LC, Krawczak FS, Gregori F, Labruna MB. Comparative susceptibility of different populations of Amblyomma sculptum to Rickettsia rickettsii. Frontiers in Physiology. 2019;**10**:653. DOI: 10.3389/fphys.2019.00653

[36] Sangioni LA, Horta MC, Vianna MCB, Gennari SM, Soares RM, Galvão MAM, et al. Ricketsial infection in animals and brazilian spotted fever endemicity. Emerging Infectious Diseases. 2005;**11**:265-269

[37] Polo G, Mera Acosta C, Labruna MB, Ferreira F. Transmission dynamics and control of *Rickettsia rickettsii* in populations of *Hydrochoerus hydrochaeris* and *Amblyomma sculptum*. PLoS Neglected Tropical Diseases. 2017;**11**:e0005613

[38] Polo G, Mera Acosta C, Labruna MB, Ferreira F, Brockmann D. Hosts mobility and spatial spread of Rickettsia rickettsii. PLoS Computational Biology 2018a;14:e1006636. pmid: 30586381

[39] Costa FB, Gerardi M, Binder LC, Benatti HR, Serpa MCA, Lopes B, et al. Rickettsia rickettsii (Rickettsiales: Rickettsiaceae) infecting Amblyomma sculptum (Acari: Ixodidae) ticks and capybaras in a Brazilian spotted fever-endemic area of Brazil. Journal of Medical Entomology. 2019;**57**(1):308-311. DOI: 10.1093/jme/tjz141

[40] Sousa KCM, Calchi AC, Herrera HM, Dumler JS, Barros-Battesti DM, Machado RZ, et al. Anaplasmataceae agents among wild mammals and ectoparasites in Brazil. Epidemiology and Infection. 2017;**145**(16):3424-3437

[41] Guedes E, Leite RC, Pacheco RC, Silveira I, Labruna MB. Rickettsia species infecting Amblyomma ticks from an area endemic for Brazilian spotted fever in Brazil. Revista Brasileira de Parasitologia Veterinária. 2011;**20**:308-311

[42] Martins TF, Onofrio VC, Barros-Battesti DM, Labruna MB. Nymphs of the genus *Amblyomma* (Acari: Ixodidae) of Brazil: Descriptions, redescriptions, and identification key. Ticks and Tick-Borne Diseases. 2010;**1**:75-99

[43] Martins TF, Barbieri AR, Costa FB, Terassini FA, Camargo LM, Peterka CR, et al. Geographical distribution of

Amblyomma cajennense (sensu lato) ticks (Parasitiformes: Ixodidae) in Brazil, with description of the nymph of A. cajennense (sensu stricto). Parasites & Vectors. 2016;**9**:1-14

[44] Dias E, Martins AV. Spotted fever in Brazil. A summary. The American Journal of Tropical Medicine and Hygiene. 1939;**19**:103-108

[45] Travassos J, Rodrigues PM, Carrijo LN. Tifo murino em São Paulo. Identificação de Rickettsia mooseri isolada de um caso humano. Memórias do Instituto Butantan. 1949;**21**:77-106

[46] Gonçalves AJR, Lopes PFA, Melo JCP, et al. Rickettsioses: A propósito de quatro casos diagnosticados no Rio de Janeiro de febre maculosa brasileira. Folha Médica. 1981;**82**:127-134

[47] Sexton DJ, Muniz M, Corey GR, et al. Brazilian spotted fever in Espirito Santo, Brazil: Description of a focus of infection in a new endemic region. The American Journal of Tropical Medicine and Hygiene. 1993;**49**:222-226. DOI: 10.4269/ajtmh.1993.49.222

[48] Queirogas EF et al. Dimensões públicas do espaço contemporâneo [tese]. Universidade de São Paulo; 2012

[49] Verdade LM, Gheler-Costa C, Penteado M, et al. The impacts of sugarcane expansion on wildlife in the state of São Paulo, Brazil. Journal of Sustainable Bioenergy Systems. 2012;**2**:138-144. DOI: 10.4236/jsbs.2012.24020

[50] Passos Nunes FB et al. The dynamics of ticks and capybaras in a residential park area in southeastern Brazil: Implications for the risk of Rickettsia rickettsii infection. Vector-Borne and Zoonotic Diseases. 2019;**19**(10):711-716. DOI: 10.1089/vbz.2019.2479

[51] Souza CE, Moraes-Filho J, Ogrzewalska M, Uchoa FC, Horta MC, Souza SS. Experimental infection of capybaras Hydrochoerus hydrochaeris by Rickettsia rickettsii and evaluation of the transmission of the infection to ticks Amblyomma cajennense. Veterinary Parasitology, Leiden. 2009. DOI: 10.1016/j.vetpar.2008.12.010

[52] Galvão MAM. Febre maculosa em Minas Gerais: Um estudo sobre a distribuição da doença no Estado e seu comportamento em área foco peri-urbano. [tesis doctoral]. Belo Horizonte: Faculdade de Medicina – UFMG; 1996.

[53] Lemos ERS et al. Primary isolation of spotted fever in the group rickettsiae from Amblyomma cooperi collected from Hydrochaeris hydrochaeris in Brazil. Memórias do Instituto Oswaldo Cruz. 1996;**91**:273-275

[54] HHB M, Colombo S, Silva MV. Spotted fever: Isolation of Rickettsia from a skin biopsy sample. Revista do Instituto de Medicina Tropical de São Paulo. 1992;**34**:37-41. DOI: 10.1590/S0036-46651992000100007

[55] de Oliveira SV, Guimarães JN, Reckziegel GC, et al. An update on the epidemiological situation of spotted fever in Brazil. Journal of Venomous Animals and Toxins including Tropical Diseases. 2016;**22**:22. DOI: 10.1186/s40409-016-0077-4

[56] Oliveira SV et al. Um caso fatal de febre maculosa brasileira em área não endêmica no Brasil: A importância de ter profissionais de saúde que entendam a doença e suas áreas de transmissão. Revista da Sociedade Brasileira de Medicina Tropical. 2016;**49**(05):653-655. DOI: 10.1590/0037-8682-0088-2016

[57] Centers for Disease Control and Prevention (CDC). Rocky Mountain

Spotted Fever (RMSF). 2022. Available from: https://www.cdc.gov/rmsf/stats/index.html

[58] European Center for Disease Prevention and Control (ECDC). Epidemiological situation of rickettsioses in EU/EFTA countries. Technical Report, 2013

[59] Favacho ARM, Rozental T, Calic SB, Scofield MAM, Lemos ERS. Fatal Brazilian spotless fever caused by Rickettsia rickettsii in a dark-skinned patient. Revista da Sociedade Brasileira de Medicina Tropical. 2011;**44**(3):395-396. DOI: 10.1590/S0037-86822011000300028

[60] Walker DH, Raoult D. Rickettsia rickettsii and other spotted fever group rickettsiae (Rocky Mountain spotted fever and other spotted fever). In: Mandell, Douglas, Bennett, editors. Principles and Practices of Infectious Diseases. 6ª edição ed. Filadélfia, PA: Churchill Livingstone; 2005. pp. 2287-2295

[61] Labruna MB, Santos FC, Ogrzewalska M, Nascimento EM, Colombo S, Marcili A, et al. Genetic identification of rickettsial isolates from fatal cases of Brazilian spotted fever and comparison with Rickettsia rickettsii isolates from the American continents. Journal of Clinical Microbiology. 2014;**52**:3788-3791. DOI: 10.1128/JCM.01914-14

[62] Angerami RN, Câmara M, Pacola MR, Rezende RCM, Duarte RMR, Nascimento EMM, et al. Features of Brazilian spotted fever in two different endemic areas in Brazil. Ticks Tick Borne Diseases. 2012;**3**(5-6):346-348. DOI: 10.1016/j.ttbdis.2012.10.010

[63] Silva N, Eremeeva ME, Rozental T, Ribeiro GS, Paddock CD, Ramos EAG, et al. Eschar-associated spotted fever rickettsiosis, Bahia, Brazil. Emerging Infectious Diseases. 2011;**17**(2):275-278. DOI: 10.3201/eid1702.100859

[64] Chapman AS, Bakken JS, Folk SM, Paddock CD, Bloch KC, Krusell A, et al. Diagnosis and management of tickborne rickettsial diseases: Rocky Mountain spotted fever, ehrlichioses, and anaplasmosis—United States: A practical guide for physicians and other health-care and public health professionals. MMWR - Recommendations and Reports. 2006;**55**:1-27

[65] Oliveira SV, Willemann MCA, Gazeta GS, Angerami RN, Gurgel-Gonçalves R. Predictive factors for fatal tick-borne spotted fever in Brazil. Zoonoses Public Health. 2017;**64**:44-50. DOI: 10.1111/zph.12345

[66] Brasil. Ministério da Saúde. Portaria N° 16, de 15 de maio de. Torna pública a decisão de incorporar a doxicilina injetável e o cloranfenicol suspensão para terapêutica da febre maculosa brasileira e outras riquetsioses no Sistema Único de Saúde – SUS. 2014. [Available in]: http://conitec.gov.br/images/incorporados/doxiciclina-e-clorafenicol-final.pdf

[67] Rocha JL, Brito JC, Nielsen R, Godinho R. Convergent evolution of increased urine-concentrating ability in desert mammals. Trends in Ecology and Evolution. 2021;**36**(7):637-650. DOI: 10.1111/mam.12244

[68] Houaiss A, Villar MS, Franco FMM. Dicionário Houaiss da Língua Portuguesa. Rio de Janeiro: Ed. Objetiva; 2004

[69] Moreira JR, Wiederhecker H, Ferraz KMPMB, Aldana-Domínguez J, Verdade LM, Macdonald DW. Capybara demographic traits. In: Moreira JR, Ferraz KMPMB, Herrera EA,

Macdonald DW, editors. Capybara: Biology, Use and Conservation of an Exceptional Neotropical Species. New York: Springer; 2012. pp. 147-167

[70] Ferraz KMPMB, Ferraz SFB, Moreira JR, Couto HT, Verdade LM. Capybara (Hydrochoerus hydrochaeris) distribution in agroecosystems: A cross-scale habitat analysis. Journal of Biogeography. 2007;**34**:223-230

[71] Verdade LM, Ferraz KMPMB. Capybaras (Hydrochoerus hydrochaeris) in an anthropogenic habitat in Southeastern Brazil. Brazilian Journal of Biology. 2006;**66**(1b):371-378. DOI: 10.1590/S1519-69842006000200019

[72] Labruna MB. Brazilian spotted fever: The role of capybaras. In: Moreira JR, KMPMB F, Herrera EA, Macdonald DW, editors. Capybara: Biology, Use and Conservation of an Exceptional Neotropical Species. 1ª ed. 2012. pp. 371-383. DOI: 10.1007/978-1-4614-4000-0_23

[73] Bratman GN, Hamilton JP, Hahn KS, Daily GC, Gross JJ. Nature experience reduces rumination and subgenual prefrontal cortex activation. Proceedings of the National Academy of Sciences of the United States of America. 2015;**112**:8567-8572. DOI: 10.1073/pnas.1510459112

[74] Díaz S, et al. Linking functional diversity and social actor strategies in a framework for interdisciplinary analysis of nature's benefits to society. Proceedings of the National Academy of Sciences of the United States of America. 2011;**108**:895-902. DOI: 10.1073/pnas.1017993108

[75] Taylor AF, Frances EK, William CS. Views of nature and self-discipline: Evidence from inner city children. Journal of Environmental Psychology. 2002;**22**(1-2):49-63

[76] Wilson EO. Biophilia and the conservation ethic. In: Kellert SR, Wilson EO, editors. The Biophilia Hypothesis. Washington DC: Island Press; 1993. pp. 31-41

[77] Zhang R, Zhang CQ, Rhodes RE. The pathways linking objectively-measured greenspace exposure and mental health: A systematic review of observational studies. Environmental Research. 2021;**198**:111-233. DOI: 10.1016/j.envres.2021.111233

[78] Liao J, Zhang B, Xia W, Cao Z, Zhang Y, Liang S, et al. Residential exposure to green space and early childhood neurodevelopment. Environment International. 2019;**128**:70-76. DOI: 10.1016/j.envint.2019.03.070

[79] Andrusaityte S, Grazuleviciene R, Dedele A, Balseviciene B. The effect of residential greenness and city park visiting habits on preschool children's mental and general health in Lithuania: A cross-sectional study. International Journal of Hygiene and Environmental Health. 2020;**223**(1):142-150. DOI: 10.1016/j.ijheh.2019.09.009

[80] Cherrie MPC, Shortt NK, Mitchell RJ, Taylor AM, Redmond P, Thompson CW, et al. Pearce Green space and cognitive ageing: A retrospective life course analysis in the Lothian Birth Cohort 1936. Social Science & Medicine. 2018;**196**:56-65

[81] Terry H et al. Nature and health. Annual Review of Public Health. 2014;**35**:207-228

[82] Medeiros B. Crianças vão ao Zoo do DF e voltam para casa infestadas de carrapatos. Metrópoles, 2018. Available from: https://www.metropoles.com/distrito-federal/criancas-vao-ao-zoo-do-df-e-voltam-para-casa-infestadas-de-carrapatos

[83] Angerami RN, Silva MV, Santos FCP, et al. Febre Maculosa Brasileira: Aspectos clínicos, epidemiológicos, diagnósticos e terapêuticos. In: Meira AM, Monti JA, Ferraz KMPMB, et al. editors. Febre Maculosa: Dinâmica da Doença, Hospedeiros e Vetores. 2013. pp. 32-51. ISBN: 97885-86481-28-4.

[84] Spolidorio MG, Labruna MB, Mantovani E, Brandão PE, Richtzenhain LJ, Yoshinari NH. Novel spotted fever group rickettsiosis, Brazil. Emerging Infectious Diseases. 2010;**16**(3):521-523

[85] Fiol FSD, Junqueira FM, Rocha MCP, Toledo MI, Barberato Filho S. A febre maculosa no Brasil. Revista Panamericana de Salud Pública. 2010;**27**:461-466

[86] Ribeiro KT, Rocha GFS, Saraiva DG, Silva AP, Vilela DAR, Lima PCS, et al. Das capivaras e carrapatos a uma proposta de comunicação e manejo no Parque Nacional da Serra do Cipó para redução de riscos à saúde. Oecol Aust. 2010;**14**(3):668-685

[87] Philippe D. Jenseits von Natur und Kultur. Berlin; 2013. p. 638. DOI: 10.15463/rec.1189729441

Section 4

Structure and Function of Vision

Chapter 11

Ommochromes of the Compound Eye of Arthropods from the Insects and Crustaceans Classes: Physicochemical Properties and Antioxidant Activity

Alexander E. Dontsov and Mikhail A. Ostrovsky

Abstract

The chapter is devoted to the study of the physicochemical properties of the ommochromes of the compound eye of arthropods. Ommochromes are the characteristic pigments of invertebrates. They are believed to function in the eyes as screening and protective pigments that protect photoreceptor cells from the damaging effects of light. Ommochromes were isolated, purified, and obtained in preparative quantities from crustaceans (Crustacea; order Decapoda) and insects (Insecta; families Stratiomyidae, Sphingidae, Blaberidae, Acrididae, and Tenebrionidae). The physicochemical properties of the isolated ommochromes were studied by absorption and fluorescence spectroscopy, electron spin resonance (ESR) and Mossbauer spectroscopy, and high-performance liquid chromatography. The antioxidant activity of ommochromes was studied by methods of inhibiting lipid peroxidation induced by reactive oxygen species and variable valence metal ions and by quenching luminol chemiluminescence. The data obtained are important both for understanding the biological functions of arthropod eye ommochromes and for the development of new pharmacological preparations based on ommochromes for the prevention and treatment of pathologies associated with the development of oxidative stress.

Keywords: crustacea, insects, eye, ommochromes, screening pigments, antioxidant activity

1. Introduction

Numerous species of animals that arose during the “Cambrian Explosion” (540–490 million years ago) had great diversity in the structure of the organs of vision. However, all this diversity can be divided into two main types: the first is the compound eye of most invertebrates (the most typical example is the eyes of arthropods), and the second is the chambered eye of vertebrates. The compound eye of arthropods consists of many small ocelli (ommatidia), each with its own photoreceptor cells (rhabdoms). The eyes contain both light-sensitive visual pigments and screening pigments. The main function of the screening pigments in the eye is light-filtering and light-absorbing. By absorbing and transmitting light in certain areas of the spectrum, preventing its reflection and scattering, screening pigments play an important role in shaping the spectral sensitivity of the eye; due to the absorption of scattered light, they determine the resolution of the eye (contrast and sharpness of the image); a number of other properties of visual perception also depend on them.

The main screening pigments of the compound eye of arthropods are organelles containing ommochromes [1]. Ommochromes are some of the main pigments of the compound eye of arthropods that protect photoreceptor cells from the damaging effects of ultraviolet light, visible light, and reactive oxygen species, both by optical shielding and by chemical neutralization of free radical products. An interesting fact of evolution: although almost all invertebrate species can synthesize and accumulate melanins, their visual organs contain mainly ommochromes, not melanins, as screening pigments.

The ommochromes were apparently first discovered by A. Johansen in 1924 in the primary and secondary pigment cells of the ommatidium of the compound eye of Drosophila, which contained two different types of pigment granules—purple-red and ocher [2]. The term "ommochromes" itself first appeared in the works of the German researcher E. Becker [3–5]. Becker proposed the general name "Ommochromes" for the pigments contained in the eyes of Drosophila and *Calliphora* flies, subdividing them into two large groups, "Ommatins" and "Ommines" [6]. Then, in the 1950s and 1960s, the molecular structure and the main physicochemical characteristics of ommochromes from a wide variety of classes and species of invertebrates were studied in detail. These numerous works were carried out by German chemists Butenandt and co-authors [1, 7, 8]. The antioxidant activity of arthropod eye ommochromes (shrimp: *Pandalus latirostris*, dragonfly: *Calopteryx splendens*, and butterfly: *Pieris brassicae*) was discovered in the 1980s [9, 10]. In 1985, it was shown also that arthropod eye ommochromes, like melanins, have a stable ESR signal with a high concentration of paramagnetic centers, which increases when exposed to ultraviolet and visible light [11].

2. Ommochromes of the compound eye of crustaceans—chemical nature, pathways of biosynthesis, and main physiochemical characteristics

2.1 Chemical structure and biosynthesis of ommochromes

In the eyes of invertebrates, ommochromes are localized in both pigment and receptor retinular cells. In addition to the eye, ommochromes have also been found in the cuticle, in excrement, and even in cells in the nervous system of insects [12]. In the cell, ommochromes are associated with specific proteins and are located in specialized organelles, ommochrome-containing granules [13]. Since the term "granules" does not imply the presence of an outer membrane, and ommochrome-containing organelles are surrounded by a membrane, it was proposed to call granules containing ommochromes ommochromasomes [14, 15]. The chemical precursors of ommochromes, especially 3-hydroxykynurenine, are thought to be transported into ommochromasomes by ATP-dependent transmembrane transporters of the ABC family [16]. Thus, in Drosophila, ommochrome precursors are transported by heterodimers of the White/Scarlet ABC transporter and its involvement in the transport of ommochrome precursors is believed to be evolutionarily ancient and widespread, particularly in insects.

Organelles containing ommochromes are inherent in almost all cells of the ommatidium. In the eyes of arthropods, these organelles are localized both in primary and secondary pigment cells and in receptor retinular cells. In shape, these are spherical formations 0.2–2 μm long surrounded by a single-layer membrane [13, 17–19], and in pigment cells, the diameter of ommochromasomes, as a rule, is much larger than in photoreceptor ones [20]. Ommochromasomes usually contain electron-dense osmiophilic material. This material likely contains complexes of ommochrome precursors and ommochromes themselves with ommochrome-binding proteins [15].

The ommochromes are generally divided into two main types—the yellow-red ommatins and the less-studied purple ommins. All ommochromes contain a phenoxazine/phenoxazone ring structure in their composition. The difference between ommatins and ommins is due to the fact that ommins presumably contain an additional phenothiazine ring (**Figure 1**).

The third type of ommochromes, ommidins, is much less common in arthropods and was found mainly in Orthoptera (Orthoptera) [21, 22]. According to their chemical structure, ommidins contain sulfur and, presumably, by analogy with pheomelanins, they are products of co-oxidation of two amino acids, in this case, tryptophan and cysteine.

Since ommochromes easily enter into various redox reactions, decarboxylation and esterification reactions, there is a large biological diversity of ommochromes in nature [15, 23]. Of the ommatins, the most common are xanthommatins, which have a characteristic phenoxazine group in their structure (**Figure 2**) [18, 24, 25]. They are especially common in the eyes of insects.

In addition to xanthommatin, other ommatins, derivatives of xanthommatin, occur in the eyes of arthropods. The most common are rhodommatin, which is a β-glycoside of dihydroxanthommatin, and ommatin D, rhodommatin sulfate. Of the ommins, ommin A is the most common. The eyes of crustaceans and most insect species contain ommins that have maximum absorption in the visible region of the spectrum around 520 nm. The ommochromes are synthesized from tryptophan via a

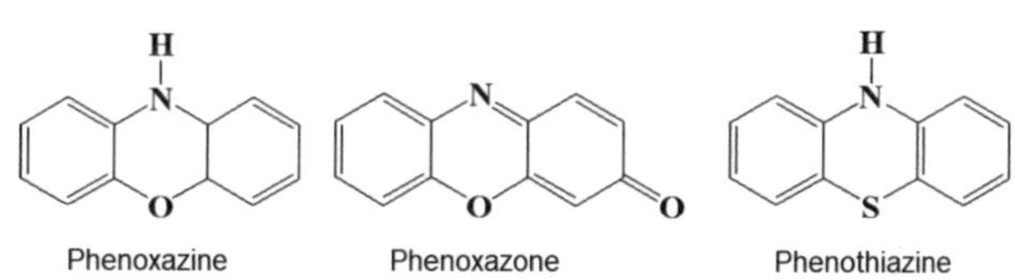

Figure 1.
Basic ring structures of ommochromes.

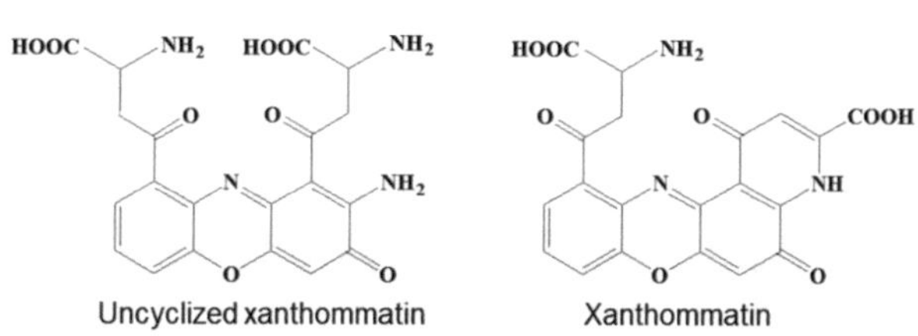

Figure 2.
Structure of xanthommatin and it uncyclized form.

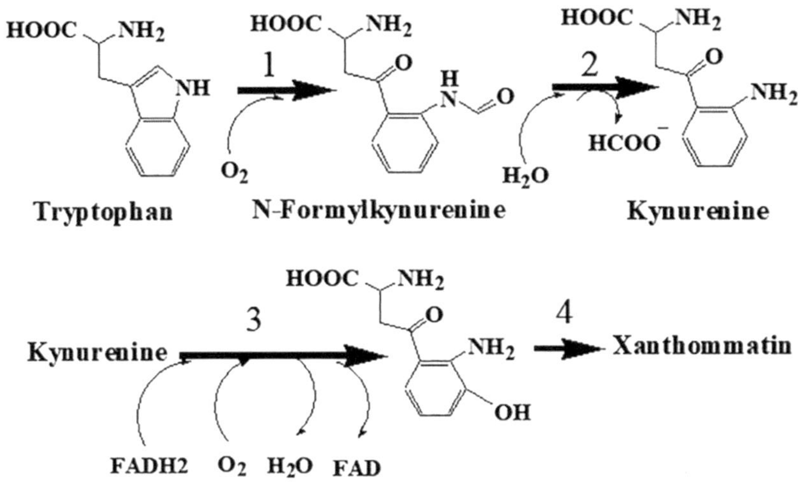

Figure 3.
Scheme of the main pathways for the biosynthesis of ommochromes. Enzymes involved in the biosynthesis of ommochromes: (1) tryptophan 2,3-dioxygenase, (2) kynurenine formamidase, (3) kynurenine 3-monooxygenase, (4) phenoxazone synthase.

kynurenine intermediate [15, 18]. **Figure 3** shows a diagram of the main pathways for the biosynthesis of ommochromes.

At the first stage of synthesis, tryptophan is oxidized to formylkynurenine. This process is catalyzed by an enzyme, tryptophan 2,3-dioxygenase (TDO) (tryptophan pyrrolase; EC 1.13.11.11). This enzyme, which was first identified in Drosophila ("vermilium" gene required for the synthesis of brown eye pigment), is a tetrameric complex containing a heme required for the oxidation of tryptophan [26, 27]. At the second stage of synthesis, kynurenine is formed. This process can be either spontaneous or catalyzed by the enzyme kynurenine formamidase (KFase; EC 3.5.1.9) [28]. Since N-formylkynurenine is known to be unstable and is rapidly converted to kynurenine in vitro, the KFase enzyme may not be required for this process. The third step in biosynthesis is the hydroxylation of kynurenine to 3-hydroxykynurenine by the enzyme kynurenine-3-monooxygenase (KMO; EC 1.14.13.9). KMO synthesis is encoded by the cinnabar gene in *Drosophila melanogaster*. KMO contains FAD as a cofactor. FAD is reduced to its active form FADH2 by NADPH. Then FADH2 oxidizes kynurenine in the presence of oxygen, which leads to the formation of 3-hydroxykynurenine [29]. In the last step, xanthommatin is formed by the condensation of two molecules of 3-hydroxykynurenine. It is still unclear whether the enzyme phenoxazone synthase (PHS; EC 1.10.3.4) is involved in this process. Thus, several works [30–33] have shown that ommochrome-containing organelles can accelerate the formation of xanthommatin in both enzymatic and non-enzymatic ways. It is assumed that the precursor of xanthommatin may be a labile uncyclized xanthommatin (**Figure 2**), which is initially formed by the condensation of two molecules of 3-hydroxykynurenine. This uncyclized xanthommatin, which has been extracted from crustaceans and insects, can spontaneously form xanthommatin at room temperature [34–37].

The processes leading to the biosynthesis of reduced xanthommatin, as well as the reactions of formation of ommatin D, rhodommatin, as well as the biosynthesis of ommins, have not yet been sufficiently studied.

2.2 Physical and chemical properties of ommochromes

2.2.1 Redox properties

Ommochromes are colored substances responsible for coloration of many invertebrates. Xanthommatin and its derivatives, such as ommatin D and decarboxylated xanthommatin, are known to determine color and its changes in arthropods [38–40]. Changes in the coloration of ommochromes are usually associated with redox transitions in the pigment molecule. In this case, the reduction leads to a shift of the absorption maximum to the longer wavelength region of the spectrum (bathochromic shift). For example, during puberty in some species of dragonflies, the body color changes from yellow to red, which is associated with the appearance of a larger amount of reduced ommochromes [41].

The chemical properties of ommochromes contribute to their participation in electron transport, in reactions with oxidizing agents, reducing agents and free radicals. It is well known [6, 42, 43] that arthropod ommochromes are easily oxidized by hydrogen peroxide or potassium superoxide. In this case, a significant shift of the absorption maximum to the shorter wavelength region (hypochromic shift) is observed. The reaction with hydrogen peroxide proceeds in at least two stages. First, the transition of ommochromes to the oxidized form occurs, which is probably followed by a gradual destruction of the pigment during prolonged incubation with an oxidizing agent, which manifests itself in a further decrease in the absorption of the pigment in the visible range [43].

Ommochromes can also act as a reducing agent in redox reactions. For example, we have previously shown that shrimp eye ommochromes easily oxidize ferrous ions to ferric ions [44]. In this work, ^{57}Fe sulfate salt was used and the Mössbauer spectra of the formed iron complexes with ommochromes were studied. The obtained gamma resonance spectra were characteristic of high-spin Fe^{3+} complexes, which indicated complex formation and simultaneous oxidation of Fe^{2+} ions by the ommochromes. The absence of relaxation spectra for Fe^{3+} – ommochrome complexes indicated their cluster nature. The Fe^{3+} ions bound in a complex with ommochromes were located close enough to each other, which ensured an effective spin-spin interaction, which led to a rapid relaxation of the electron spin and "collapse" of the magnetic hyperfine structure [44].

The coordination of iron ions apparently occurs with the carboxyl, amino and imino groups of the ommochrome molecules. This can be represented within the structure of xanthommatin by the formation of a six-membered metallocycle with a system of conjugated bonds due to the coordination of Fe^{3+} with the imine nitrogen atom (1) and oxygen atom (2) of the neighboring ring according to the scheme (**Figure 4**):

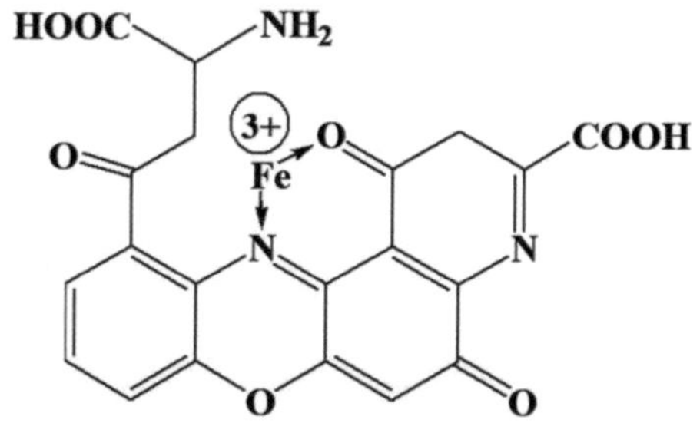

Figure 4.
Hypothetical structure of iron-xanthommatin complex.

This exceptionally efficient binding of Fe^{2+} by ommochromes (with subsequent oxidation to Fe^{3+}) into prooxidant inactive complexes may be one of the important mechanisms of the antioxidant action of shielding pigments in the arthropod eye [45, 46].

2.2.2 Free-radical properties of ommochromes

Previously, we showed that the ommochromes of the eyes of dragonflies and shrimps, similarly to melanins, exhibit a stable paramagnetic resonance signal [11]. These results were confirmed for the eye ommochromes of insects of various families [43]. All the studied ommochromes had a pronounced singlet electron spin resonance (ESR) signal with values of g factors close to the g factor of a free electron (2.0045–2.0048) and a fairly high concentration of unpaired electrons (> 10^{17} spin/g dry weight). **Figure 5A** shows the ESR spectra of ommochromes for 4 insect species.

Irradiation of ommochromes with ultraviolet or visible region of the spectrum at liquid nitrogen temperature leads to a significant increase in the amplitude of the ESR signal (up to 5–9 times greater than the original dark signal) without changing the nature of the signal. The signal amplitude reaches saturation after approximately 40–50 min of exposure to light at the temperature of liquid nitrogen. When the light is turned off and the temperature is 1–3°C, the intensity of the ESR signal returns to its original level within 30–60 seconds. The ESR signal of ommochromes also turned out to be sensitive to the action of hydrogen peroxide (**Figure 5B**, curves 1 and 2). Oxidative destruction of the ommochromes by hydrogen peroxide led to a sharp drop in the ESR signal and, ultimately, to a complete loss of paramagnetism, which is probably due to the destruction of the phenoxazine ring in the structure of ommochrome molecules [47], which initially exhibits free radical properties. Destruction of ommochromes from the Drosophila eye with hydrogen peroxide has been shown previously [6].

The high concentration of stable free radical centers makes it possible to consider ommochromes as scavengers of active free radicals. The value of g factor of ommochromes, which is in the range between 2.004 and 2.005, is typical for phenoxy radicals [48]. It is known that the intermediates of phenoxazine, which is part of the structure of the ommochrome molecule, exhibit a stable ESR signal [49, 50] and can, apparently, cause the ESR signal that we found in ommochromes of insects. Moreover, it is possible that phenoxazine determines the antiradical activity of ommochromes [47].

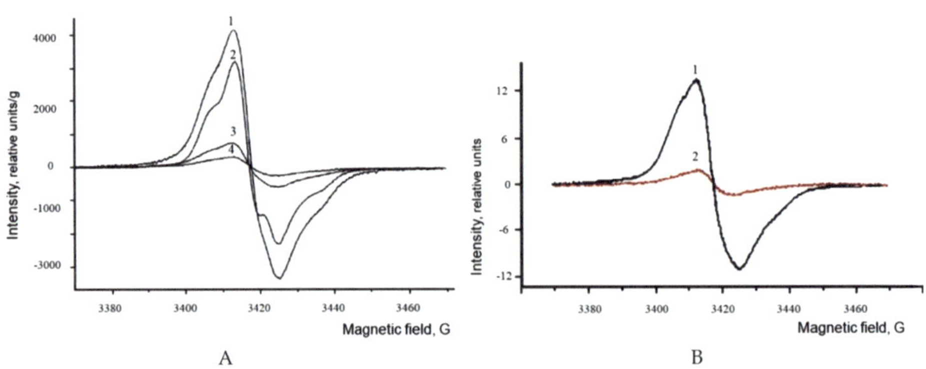

Figure 5.
ESR spectra of ommochromes. (A) ESR spectra of ommochromes of the black soldier fly (1), butterflies tobacco hawk moth (2), marbled cockroach (3), and desert locust (4). (B) Oxidation of ommochromes with hydrogen peroxide (curve 2) result in drop of ESR signal; (1) original spectrum.

2.2.3 Antioxidant activity of ommochromes

The ommochromes of all studied arthropod species exhibit high antioxidant activity by inhibiting the lipid peroxidation reaction induced by various reactive oxygen species and the action of ultraviolet and visible radiation [15, 43, 45, 51, 52]. **Figure 6** demonstrates the inhibitory effect of fly eye ommochromes on the peroxidation of the outer segments of photoreceptors (POS) of the bull eye (**Figure 6A**) and shrimp eye ommochromes on the process of peroxidation of POS of the frog eye (**Figure 6B**). Fly eye ommochromes at a concentration of 350 μg/mL reduced the POS peroxidation rate by more than threefold. The antioxidant activity of ommochromes was most pronounced when the process of POS peroxidation was induced under conditions of hyperoxia. Hyperoxia was induced by intense bubbling of pure oxygen in the POS suspension (**Figure 6B**).

It is important to note that in this case, the oxidation conditions were closer to those in vivo. It can be seen that under hyperoxia there is an increase in the rate of accumulation of TBA-reactive products in the POS suspension by about twofold. The presence of ommochromes at a low concentration (40 μg/mL) completely suppresses this effect. The antioxidant activity of ommochromes may be associated with the removal of free radicals due to the antiradical properties of these molecules [15, 52–54].

The redox balance in the cell is determined by the ratio of free radicals and antiradical molecules. When the concentration of free radicals is exceeded, oxidative stress occurs. The presence of antiradical molecules capable of neutralizing free radicals contributes to the control of oxidative stress. Apparently, there are two mechanisms of interaction of ommochromes with free radicals—electron transfer and hydrogen atom transfer [54]. It is known that ommochromes easily neutralize superoxide anion radicals [45, 52, 55, 56]. We have previously shown that shrimp eye ommochromes enhance the breakdown of superoxide radicals in both aqueous and anhydrous media (**Figure** 7) [55].

Superoxide anion radicals obtained by electrochemical reduction of oxygen on a mercury cathode (at a current of 2 mA) in anhydrous medium (a solution of tetrabutylammonium iodide in dimethylformamide) slowly decompose by reacting with each other (superoxide concentration was measured by the degree of reduction of nitro blue tetrazolium or cytochrome c). The addition of shrimp eye ommochromes

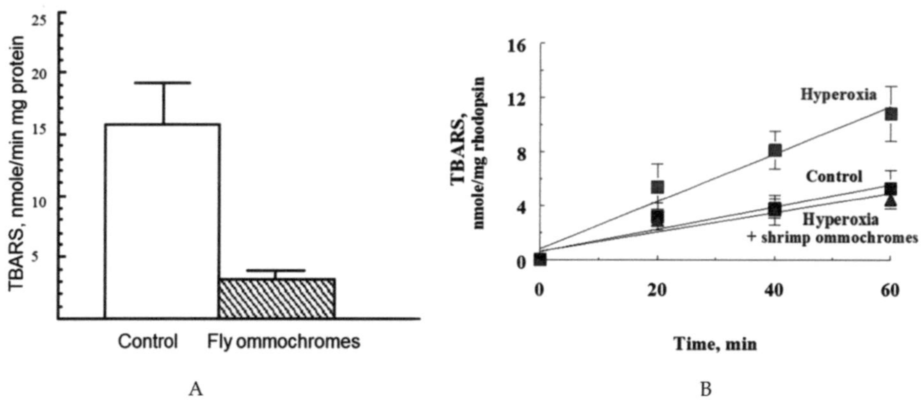

Figure 6.
Inhibitory effect of fly ommochromes (A) and shrimp ommochromes (B) on lipid peroxidation. (A) Inhibition of POS peroxidation induced by Fe^{2+}-ascorbic acid. (B) Inhibition of POS peroxidation induced by hyperoxia.

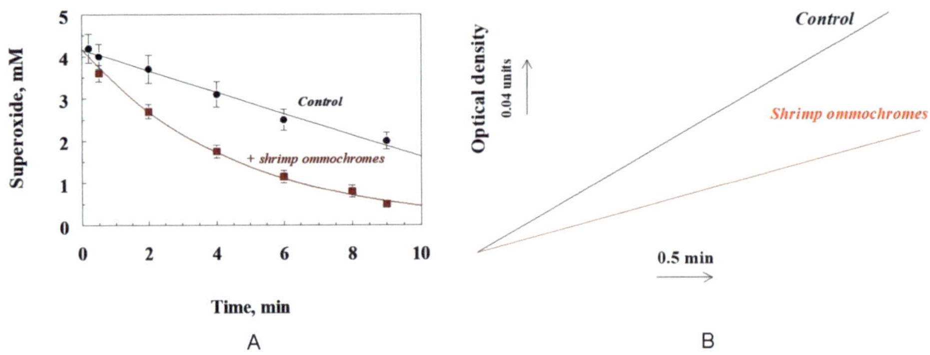

Figure 7.
Acceleration of the decomposition of superoxide radicals by shrimp eye ommochromes in anhydrous (A) and in aqueous (B) media.

(170 μg/ml) led to a significant acceleration of the process of decomposition of superoxide radicals (**Figure 7A**). Ommochromes also accelerate the destruction of superoxide radicals in an aqueous medium (**Figure 7B**). Superoxide radicals were generated in the oxidation of xanthine to uric acid catalyzed by the enzyme xanthine oxidase. In the presence of shrimp ommochromes (50 μg/mL) a significant decrease in the rate of adrenaline oxidation to adrenochrome (480 nm) by superoxide radicals is observed [55].

The antiradical activity of ommochromes is apparently due to the presence of phenoxazine and phenothiazine rings in their structure. Phenoxazine-based compounds are known to be good antioxidants [47]. The antiradical properties of ommochromes can be directly linked to the N–H group of the phenoxazine/phenothiazine ring, which means that only reduced ommochromes can act as powerful antiradical and antioxidant molecules in cells [47]. It has been shown that phenoxazines and phenothiazines under in vivo conditions are strong inhibitors of autoxidation and ferroptosis (iron-dependent oxidative stress) due to the neutralization of lipid radicals [57]. Recent studies have shown that phenoxazine-based compounds can be used to protect against oxidative stress [47, 54, 57].

Thus, arthropod ommochromes have a pronounced antioxidant activity, inhibiting peroxidation processes even at relatively low concentrations. The antioxidant activity of insect ommochromes was comparable to that of natural melanins [42] and synthetic antioxidants of the oxypyridine series [58]. On this basis, we can confidently assume that the ommochromes of the arthropod eye at physiological concentrations have antioxidant effects.

2.2.4 Light sensitivity of ommochromes

The ommatins are known to be sensitive to irradiation. They readily undergo methylation, methoxylation, and decarboxylation under the influence of light in acidified methanol [15, 59–61]. Some studies have reported that during the incubation of the extracted ommochromes in acidified methanol for several hours at room temperature, the formation of ommochromes was observed that differed in spectral characteristics from the original pigment [37]. Under the influence of light on the ommochromes, the generation of superoxide anions is observed [62].

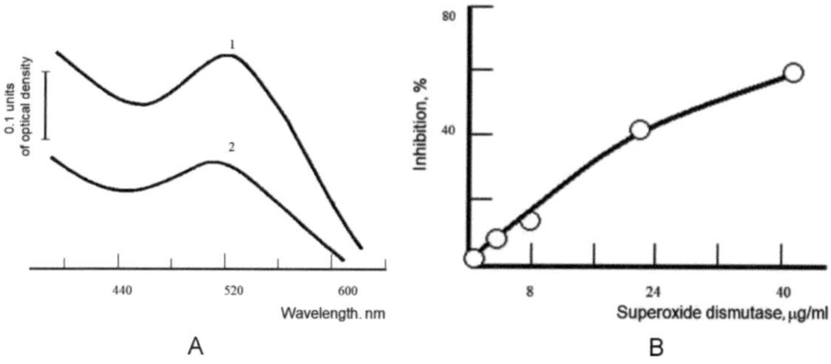

Figure 8.
Light-dependent reduction of nitro blue tetrazolium by shrimp eye ommochromes (A) and the inhibitory effect of superoxide dismutase (B).

In particular, it was shown that illumination of a system containing a suspension of shrimp eye ommochromes in the presence of detergent cetyltrimethylammonium bromide (CTAB) and nitro blue tetrazolium reduces the latter with the formation of formazan (**Figure 8**).

Illumination of the systems in control samples containing either only nitroblue tetrazolium and CTAB, or only ommochromes in the presence of CTAB did not lead to any spectral changes. The ommochromes reduced nitro blue tetrazolium in the presence of a detergent and in the dark, but the reaction rate was an order of magnitude slower than in the case of illumination. Photogeneration of superoxide radicals was observed when ommochromes were illuminated with both ultraviolet and visible light. The superoxide photogeneration process was inhibited by superoxide dismutase (**Figure 8B**). The concentration of superoxide dismutase causing 50% inhibition was 25 µg/ml.

Thus, arthropod eye ommochromes are capable of photogeneration of the superoxide radicals. When illuminated, ommochromes are able to reduce oxygen to its anion-radical form. This process, apparently, is inherent only in the reduced form of ommochromes, since oxidized ommochromes reduce NBT with much less efficiency under illumination. The physiological significance of this process is unclear. On the one hand, it is possible that this process proceeds in vivo at an insignificant rate and does not pose a risk for cell photodamage. On the other hand, it is known that oxidized phenoxazine derivatives can be reduced by glutathione during irradiation [63]. Superoxide anion radicals are an intermediate product of this reaction. Based on the structural similarity of ommochromes to phenoxazine derivatives, it can be assumed that they can also be reduced in light. Moreover, the reduced form of ommochromes can catalyze this process. Such a mechanism in the tissues of the eyes of invertebrates would be particularly effective, since oxidized ommochromes formed in the course of interaction, for example with superoxide, would be constantly renewed in light.

3. The main functions of ommochromes in the eyes of arthropods

It is generally accepted that ommochromes in the cells of the compound eye of arthropods perform the function of optical shielding of light-sensitive elements of retinular cells and individual ommatidia from each other, as well as the function of regulation of spectral sensitivity of photoreceptors [23]. Recently, more and more facts have appeared showing that ommochromes, along with carotenoids that quench singlet oxygen, can protect eye cells from oxidative stress.

3.1 Optical screening function

In light, the pigment organelles of photoreceptors and pigment cells change their position, tending to group around the rhabdom (**Figure 9**). During dark adaptation, ommochromasomes migrate away from the light-sensitive rhabdom, opening it to light. During light adaptation, they migrate back to the light-sensitive rhabdom, protecting it from excess light. Such retinomotor activity is one of the mechanisms of light and dark adaptation of the compound eye of arthropods.

Another striking example of the influence of the retinomotor effect on visual acuity is the highly developed compound eye of insects, such as the superposition eye of Diptera. The rhabdom in these eyes acts as a light guide or optical fiber. The position of the screening pigment granules that surround the microvilli of the rhabdomer containing rhodopsin changes during the daily cycle. During daylight hours, the granules surround the microvilli in such a way that the light is concentrated on them and not scattered. Therefore, dipterous day vision is highly acuity. In the dark time of the day, the granules migrate from the rhabdomers, the light is scattered, and visual acuity decreases, but the sensitivity of the eye increases.

3.2 Adjustment of the spectral sensitivity of the eye

Screening pigment-containing organelles can regulate the spectral sensitivity of the arthropod compound eye. One example is the adjustment of the spectral sensitivity of the fly's compound eye. Normally, there are two types of screening pigments in the eyes of flies: red ommochromes and yellowish pterins. First ones absorb light in the visible range, passing it only in the red region of the spectrum, the second ones absorb mainly in the ultraviolet region. Therefore, all light passing to the rhabdom through its own facet, that is, unshielded, is absorbed by the visual pigment rhodopsin with $\lambda_{max} \approx 500$ nm. The light passing to the rhabdom from the side, through the neighboring facets, is filtered by ommochromasomes. As a result, only red light reaches the rhabdom. Thus, the fly's eye turns out to be endowed with a peculiar additional red-sensitive (λ_{max} = 620 nm) light detector of a "filter" nature. Combined

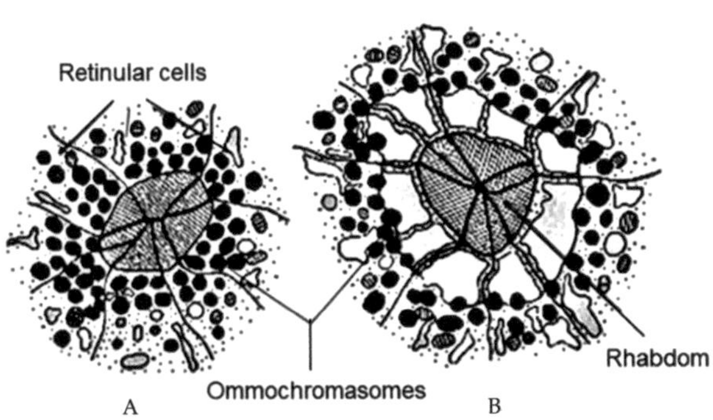

Figure 9.
Ommochromasomes migrate during light (A) and dark adaptation (B).

with the main green-sensitive ($\lambda_{max} \approx 500$ nm) light detector, this additional "filter" light detector, based on the same rhodopsin, allows flies to distinguish colors [64–66].

Another example of adjusting the spectral sensitivity of the compound eye is associated with the study of screening pigments in crustaceans—glacial-relict shrimps of the *Mysis* genus. Subspecies of the relict shrimp *Mysis relicta* in the postglacial period (about 10,000 years ago) were in different lighting conditions. It was shown that in the subspecies inhabiting the slightly saline waters of the Baltic Sea at shallow depths, the maxima of the spectral eye sensitivity curves, as a rule, are shifted by 20–30 nm to the short-wavelength region of the spectrum compared to subspecies of *M. relicta* inhabiting in fresh lake waters [67]. Studies have shown that the lake shrimp *M. relicta*, living at great depths in a brown-red light environment, has a long-wave spectral sensitivity with a maximum of about 600 nm. In contrast, the sea shrimp *M. relicta*, inhabiting at a shallow depth of the Baltic Sea, has a shorter wavelength spectral sensitivity with a maximum of about 570 nm. It was shown [68, 69] that both populations contain two types of rhodopsins in two types of rhabdoms, one with an absorption maximum at 525–530 nm, and the other with an absorption maximum at 565–570 nm. However, in the lake population of shrimp, the content of longer-wavelength rhodopsin significantly predominates, while in the sea population, on the contrary, the content of shorter-wavelength rhodopsin significantly predominates [70, 71]. It is important to emphasize that both in the sea and lake populations of *M. relicta*, the spectral sensitivity of the eye is noticeably shifted to the long-wavelength region of the spectrum with respect to the absorption spectra of their visual pigments. Using microspectrophotometry, we studied the absorption spectra of pigment-containing organelles in the eyes of both populations of *M. relicta* shrimp [67, 69]. Most pigment organelles had absorption spectra in the blue region of the spectrum, which is typical for xanthommatins (maximum at 455 nm). A smaller number of pigment organelles had absorption spectra characteristic of a mixture of ommochromes—ommatins and ommins (unpronounced maxima at 440 and 555 nm) (**Figure 10A** and **B**).

In addition, in the eyes of sea mysids, but not lake, populations were recorded in a small number of spectra characteristic of ommins (maximum about 580 nm).

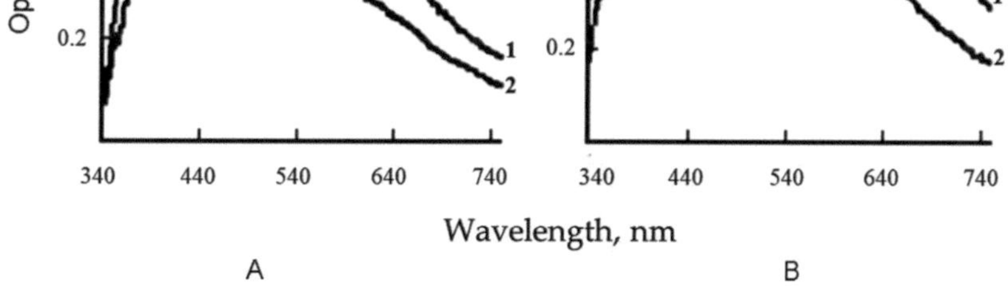

Figure 10.
Absorption spectra of eye ommochromasomes of Sea (1) and Lake (2) Mysis relicta populations. (A) Ommochromasomes with xanthommatin-like spectra; (B) Ommochromasomes with ommatin-ommin-like spectra.

Calculations of the spectral sensitivity of the eye of *M. relicta*, formed by the absorption spectra of visual pigments, taking into account the filtering of light by ommochrome-containing organelles, made it possible to construct a spectral sensitivity curve completely identical to the real spectral sensitivity of the eye of *M. relicta*. Thus, the spectral features of ommochromes in the eyes of the lake mysid population led to a shift of the maximum spectral sensitivity of the eye by 30 nm to the long wavelength region of the spectrum, which is very important for vision at great depths with low illumination.

3.3 Oxidative stress protection function

It is known that the lake population of relict shrimp *M. relicta* is much more sensitive to the damaging effects of light compared to the sea population—even moderate illumination of animals, for example, on the surface of the water in the evening in the absence of sun, leads to complete loss of vision [72]. It was suggested that one of the possible reasons for the high sensitivity of the lake shrimp population to light may be associated with a lower level of protection of eye cells from the action of free radical products induced by irradiation. This means that the system of antioxidant defense in the shrimp of the lake population is significantly lower than in the shrimp of the sea population. Indeed, a study of the comparative activity of antioxidant systems and their components in the eyes of two *M. relicta* shrimp populations showed that the sea shrimp population has a significantly more active antioxidant defense system compared to the lake population [73]. As can be seen from **Figure 11**, the homogenates obtained from the eyes of the shrimp of the lake population practically did not inhibit the process of accumulation of hydroperoxides during oxidation of cardiolipin liposomes, while the homogenates obtained from the eyes of the shrimp of the sea population were active in this respect.

The time interval required to reach the maximum concentration of hydroperoxides in the presence of the homogenate of the eyes of shrimp from the sea population was almost three times longer than in the control. At the same time, during the initial period of the reaction (up to 20 min), the content of hydroperoxides in the liposome/homogenate system of the lake population was more than 4 times higher than that for the liposome/homogenate system of the sea population. This means that the shrimp of the sea population have a higher activity of antioxidant defense systems than

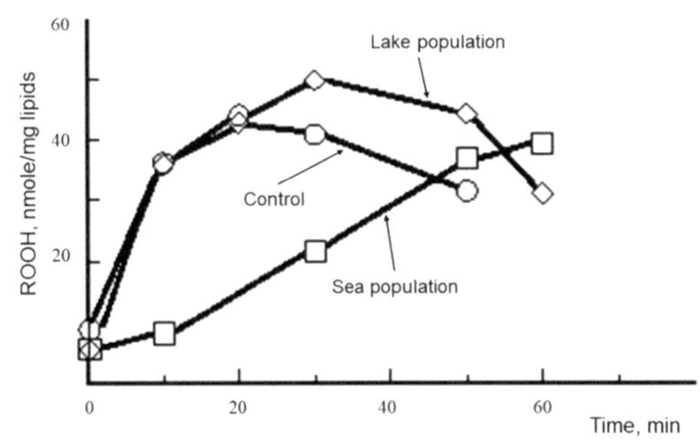

Figure 11.
Effect of eye homogenates shrimp Mysis relicta on Fe^{2+}-ascorbate-induced peroxidation of cardiolipin liposomes.

Population *M. relicta*	Eyes number	Ommochrome content	
		μg of dry weight per eye	μg of dry weight per mg protein
Lake population	150	14.3 ± 1.8	90.5 ± 11.8
Sea population	150	39.3 ± 2.9	218.6 ± 17.5

Table 1.
The content of ommochromes in the eyes of marine and lake shrimp Mysis relicta populations.

the shrimp of the lake population. The study of the content of the main systems of antioxidant defense—the activity of antioxidant enzymes (superoxide dismutase and glutathione peroxidase), as well as the concentration of α-tocopherol in the homogenates of both types of eyes, showed no significant difference for the homogenates of sea and lake shrimp populations [74]. At the same time, analysis of homogenates for the content of ommochromes showed that the concentration of ommochromes in the homogenate of the eyes of shrimp in the sea population was almost 2.5 times higher than in the lake population (**Table 1**).

Table 1 shows that the concentration of ommochromes in the eyes of sea population shrimp reaches almost 40 μg per eye. If we assume that the volume of one eye does not exceed 0.5 μl, then the concentration of ommochromes in them reaches 80 mg of dry weight per 1 ml. In shrimp of the lake population, the low concentration of ommochromes may be associated with the loss of pigments in the process of adaptation to a deep-sea lifestyle in very low light conditions. At the same time, the ommochromes of sea and lake populations are identical in their spectral and antioxidant characteristics. These data indicate that the higher antioxidant activity of the eye homogenate of shrimp from the sea population can indeed be associated with a higher content of ommochromes in them.

Another example of the antioxidant function of ommochromes is related to the suppression of oxidative stress in triatomine bugs (Hemiptera: Reduvidae). For example, the action of ultraviolet radiation on red-eyed mutants bugs lacking ommochromes led to severe oxidative stress and damage to ommatidia, especially in the case of abundant blood feeding, when a high concentration of heme, a powerful generator of reactive oxygen species, is formed in the body [51]. While in wild-type bugs, exposure to ultraviolet caused an increase in the intensity of the synthesis of ommochromes and protection of eye structures from light and oxidation.

4. Conclusion

Ommochromes, which are found in large numbers in the cells of the compound eye of arthropods, are necessary for these animals to perform a wide variety of functions. These are optical shielding of individual ommatidia from each other, increasing the resolution of the eye, expanding the range of color perception, and regulating the spectral sensitivity of photoreceptors. In addition, ommochromes can protect the eyes of arthropods from oxidative stress caused both by ultraviolet or intense visible light irradiation and by excessive generation of reactive oxygen species. But ommochromes may be necessary not only for arthropods. Recently, there have been works pointing to the antimicrobial, antifungal, and antiglycation properties of ommochromes [75–77]. Due to their biological activity, ommochromes can be promising pharmacological preparations for prevention and treatment of pathologies associated with the oxidative stress development.

Acknowledgements

This research was funded by the Ministry of Science and Higher Education of the Russian Federation, grant number 075-15-2020-773. The authors are grateful to the Russian Academy of Sciences and the Tvärminne Zoological Station of the Helsinki University for many years of assistance in carrying out these studies.

Conflict of interest

The authors declare no conflict of interest.

Author details

Alexander E. Dontsov and Mikhail A. Ostrovsky*
Emanuel Institute of Biochemical Physics, Russian Academy of Sciences; Koltzov Institute of Developmental Biology, Russian Academy of Sciences, Moscow, Russia

*Address all correspondence to: ostrovsky3535@mail.ru

References

[1] Butenandt A, Schafer W. Ommochromes. In: Gore TS, Joshi BS, Sunthankar SV, Tilak BD, editors. Recent Progress in the Chemistry of Natural and Synthetic Coloring Matters and Related Fields. New York, NY: Academic Press; 1962. pp. 13-34

[2] Johannsen AO. Eye structure in normal and eye-mutant *Drosophila*. Journal of Morphology. 1924;**39**:337-350

[3] Becker E. Über die Natur des Augenpigments von Ephestia kuhniella und seinen Vergleich mit den Augenpigmenten anderer Insekten. Biologisches Zentralblatt. 1939;**59**:597-627

[4] Becker E. Die pigmente der ommin- und ommatingruppe, eine neue klasse von naturfarbstoffen. Die Naturwissenschaften. 1941;**29**:237-238

[5] Becker E. Über Eigenschaften, Verbreitung und die genetischent-wicklungsphysiologische Bedeutung der Pigmente der Ommatin-und Ommingruppe (Ommochrome) bei den Arthropoden. Molecular & General Genetics. 1942;**80**:157-204

[6] Ephrussi B, Herold JL. Studies of eye pigments of Drosophila. Methods of extraction and quantitative estimation of the pigment components. Genetics. 1944;**29**:148-175

[7] Butenandt A, Biekert E, Koga N, Traub P. Über Ommochrome, XXI. Konstitution und Synthese des Ommatins D. Hoppe-Seyler's Zeitschrift für Physiologische Chemie. 1960;**321**:258-275

[8] Butenandt A, Biekert E, Kubler H, Linzen B, Traub P. Über Ommochrome, XXII. Konstitution und Synthese des Rhodommatins. Hoppe-Seyler's Zeitschrift für Physiologische Chemie. 1963;**334**:71-83

[9] Dontsov AE. Antioxidant function of ommochromes of the shrimp *Pandalus latirostris*. Zhurnal Evoliutsionnoĭ Biokhimii i Fiziologii. 1981;**17**:53-56. (Article in Russian)

[10] Pustynnikov MG, Dontsov AE. Inhibition of UV-induced accumulation of lipid peroxides by melanins and ommochromes. Biokhimiia (Moscow). 1988;**53**:1117-1120. PMID: 3179360. (Article in Russian)

[11] Dontsov AE, Mordvintsev PI, Lapina VA. Dark and light-induced ESR signals of the ommochromes of the invertebrate eye. Biophysics (Moscow). 1985;**30**:6-8. (Article in Russian)

[12] Sawada H, Nakagoshi M, Mase K, Yamamoto T. Occurrence of ommochrome-containing pigment granules in the central nervous system of silkworm, *Bombyx mori*. Comparative Biochemistry and Physiology. B. 2000;**125**:421-428. DOI: 10.1016/s0305-0491(99)00179-0

[13] Kayser H. Pigments. In: Kerkut GA, Gilbert LI, editors. Comprehensive Insect Physiology, Biochemistry, and Pharmacology. Vol. 10. New York: Pergamon Press; 1985. pp. 368-415

[14] Needham AE. The significance of zoochromes (Zoophysiology and ecology 3). Berlin-Heidelberg: Springer; 1974. p. 429

[15] Figon F, Casas J. Ommochromes in invertebrates: biochemistry and cell biology. Biological Reviews. 2019;**94**: 156-183. DOI: 10.1111/brv.12441

[16] Mackenzie SM, Howells AJ, Cox GB, Ewart GD. Sub-cellular localisation of the white/scarlet ABC transporter to pigment granule membranes within the compound eye of Drosophila melanogaster. Genetica. 2000;**108**:239-252. DOI: 10.1023/a:1004115718597

[17] Shoup JR. The development of pigment granules in the eyes of wild type and mutant *Drosophila melanogaster*. The Journal of Cell Biology. 1966;**29**:223-249. DOI: 10.1083/jcb.29.2.223

[18] Linzen B. The Tryptophan→ Ommochrome pathway in insects. In: Treherne JE, Berridge MJ, Wigglesworth VB, editors. Advances in Insect Physiology. Vol. 10. New York: Acad. Press; 1974. pp. 117-246

[19] Stark WS, Sapp R. Eye color pigment granules in wild-type and mutant *Drosophila melanogaster*. Canadian Journal of Zoology. 1988;**66**:1301-1308. DOI: 10.1139/z88-191

[20] Gribakin FG, Chesnokova EG. The use of eye mutants to study the physiology of insects. Uspekhi Sovremennoĭ Biologii. 1984;**97**:69-82. (Article in Russian)

[21] Über LB, Ommochrome XXIII. Ommidine -ein neuer typus von ommochromen aus orthopteran. Zeitschrift für Naturforschung. Teil B. 1966;**21**:1038-1047. DOI: 10.1515/znb-1966-1107

[22] Pinamonti S, Chiarelli-Alvisi G, Colombo G. The Ommochromes of the Eyes and the Integument of Albino and Pigmented Locusts, *SchistocercaGregaria* Forsk (Orthoptera). The Italian Journal of Zoology. 1973;**40**:41-54. DOI: 10.1080/11250007309427480

[23] Britton G. The Biochemistry of Natural Pigments. Cambridge: Cambridge University Press; 1983. p. 422

[24] Strother GK. Absorption of *Musca domestica* screening pigment. The Journal of General Physiology. 1966;**49**:1087-1088. DOI: 10.1085/jgp.49.5.1087

[25] Langer H. Properties and functions of screening pigments in insect eyes. In: Snyder AW, Menzel R, editors. Photoreceptor Optics. Berlin: Springer; 1975. pp. 429-455

[26] Capece L, Lewis-Ballester A, Batabyal D, Di Russo N, Yeh SR, Estrin DA, et al. The first step of the deoxygenation reaction carried out by tryptophan dioxygenase and indoleamine 2,3-dioxygenase as revealed by quantum mechanical/molecular mechanical studies. Journal of Biological Inorganic Chemistry. 2010;**15**:811-823. DOI: 10.1007/s00775-010-0646-x

[27] Huang W, Gong Z, Li J, Ding J. Crystal structure of *Drosophila melanogaster* tryptophan 2,3-dioxygenase reveals insights into substrate recognition and catalytic mechanism. Journal of Structural Biology. 2013;**181**:291-299. DOI: 10.1016/j.jsb.2013.01.002

[28] Moore GP, Sullivan DT. Biochemical and genetic characterization of kynurenine formamidase from *Drosophila melanogaster*. Biochemical Genetics. 1978;**16**:619-634. DOI: 10.1007/BF00484718

[29] Smith JR, Jamie JF, Guillemin GJ. Kynurenine-3-monooxygenase: a review of structure, mechanism, and inhibitors. Drug Discovery Today. 2016;**21**:315-324. DOI: 10.1016/j.drudis.2015.11.001

[30] Phillips JP, Forrest HS. Terminal synthesis of xanthommatin in *Drosophila melanogaster*. II. Enzymatic formation of the phenoxazinone nucleus. Biochemical Genetics. 1970;**4**:489-498. DOI: 10.1007/BF00486599

[31] Phillips JP, Forrest HS, Kulkarni AD. Terminal synthesis of xanthommatin in *Drosophila melanogaster*. III. Mutational pleiotropy and pigment granule association of phenoxazinone synthetase. Genetics. 1973;**73**:45-56. DOI: 10.1093/genetics/73.1.45

[32] Yamamoto M, Howells AJ, Ryall RL. The ommochrome biosynthetic pathway in *Drosophila melanogaster*: the head particulate phenoxazinone synthase and the developmental onset of xanthommatin synthesis. Biochemical Genetics. 1976;**14**:1077-1090. DOI: 10.1007/BF00485139

[33] Rasgon JL, Scott TW. Crimson: a novel sex-linked eye color mutant of *Culex pipiens* L. (Diptera: Culicidae). Journal of Medical Entomology 2004; 41: 385-391.10.1603/0022-2585-41.3.385.

[34] Bolognese A, Scherillo G. Occurrence and characterization of a labile xanthommatin precursor in some invertebrates. Experientia. 1974;**30**:225-226. DOI: 10.1007/BF01934793

[35] Iwahashi H, Ishii T. Detection of the oxidative products of 3-hydroxykynurenine using high-performance liquid chromatography–electrochemical detection–ultraviolet absorption detection–electron spin resonance spectrometry and high-performance liquid chromatography–electrochemical detection–ultraviolet absorption detection–mass spectrometry. Journal of Chromatography. A. 1997;**773**:23-31. DOI: 10.1016/S0021-9673(97)00271-9

[36] Zhuravlev AV, Vetrovoy OV, Savvateeva-Popova EV. Enzymatic and non-enzymatic pathways of kynurenines' dimerization: the molecular factors for oxidative stress development. PLoS Computational Biology. 2018;**14**:e1006672. DOI: 10.1371/journal.pcbi.1006672

[37] Figon F, Munsch T, Croix C, Viaud-Massuard M-C, Lanoue A, Casas J. Uncyclized xanthommatin is a key ommochrome intermediate in invertebrate coloration. Insect Biochemistry and Molecular Biology. 2020;**124**:103403. DOI: 10.1016/j.ibmb.2020.103403

[38] Riou M, Christides J-P. Cryptic color change in a crab spider (Misumena vatia): Identification and quantification of precursors and ommochrome pigments by HPLC. Journal of Chemical Ecology. 2010;**36**:412-423. DOI: 10.1007/s10886-010-9765-7

[39] Stavenga DG, Leertouwer HL, Wilts BD. Coloration principles of nymphaline butterflies–thin films, melanin, ommochromes and wing scale stacking. The Journal of Experimental Biology. 2014;**217**:2171-2180. DOI: 10.1242/jeb.098673

[40] Panettieri S, Gjinaj E, John G, Lohman DJ. Different ommochrome pigment mixtures enable sexuall dimorphic Batesian mimicry in disjunct populations of the common palmfly butterfly, Elymnias hypermnestra. PLoS One. 2018;**13**:e0202465. DOI: 10.1371/journal.pone.0202465

[41] Futahashi R, Kurita R, Mano H, Fukatsu T. Redox alters yellow dragonflies into red. Proceedings of the National Academy of Sciences. 2012;**109**:12626-12631. DOI: 10.1073/pnas.1207114109

[42] Ushakova N, Dontsov A, Sakina N, Bastrakov A, Ostrovsky M. Antioxidative Properties of Melanins and Ommochromes from Black Soldier Fly Hermetia illucens. Biomolecules. 2019;**9**:408. DOI: 10.3390/biom9090408

[43] Dontsov AE, Sakina NL, Yakovleva MA, Bastrakov AI,

Bastrakova IG, Zagorinsky AA, et al. Ommochromes from the Compound Eyes of Insects: Physicochemical Properties and Antioxidant Activity. Biochemistry (Moscow). 2020;**85**:668-678. DOI: 10.1134/S0006297920060048

[44] Bagirov RM, Stukan RA, Dontsov AE, Ostrovsky MA, Lapina VA. Investigation of the binding of iron ions by ommochromes of the eye of invertebrates using gamma-resonance spectroscopy. Biophysics (Moscow). 1986;**31**:1017-1022. (Article in Russian)

[45] Ostrovsky MA, Sakina NL, Dontsov AE. Antioxidative role of eye screening pigments. Vision Research. 1987;**27**:893-899. DOI: 10.1016/0042-6989(87)90005-8

[46] Dontsov AE, Ostrovsky MA. The antioxidant role of shielding eye pigments – melanins and ommochromes, and physicochemical mechanism of their action. In: Varfolomeev SD, Burlakova EB, editors. Chemical and Biological Kinetics New Horizons. Vol. 2. Danvers: CRC Press; 2005. pp. 133-150

[47] Farmer LA, Haidasz EA, Griesser M, Pratt DA. Phenoxazine: a privileged scaffold for radical trapping antioxidants. The Journal of Organic Chemistry. 2017;**82**:10523-10536. DOI: 10.1021/acs.joc.7b02025

[48] Bolton JR. Experimental aspects of biological electron spin resonance studies. In: Swartz HM, Bolton JR, Borg DC, editors. Biological Application of Electron Spin Resonance. Wiley (Interscience): New York, NY; 1972. pp. 64-118

[49] Lhoste J-M, Haug A, Ptak M. Electron paramagnetic resonance studies of photoselected triplet molecules. I. Phenoxazine. The Journal of Chemical Physics. 1966;**44**:648-654. DOI: 10.1063/1.1726739

[50] Bolognese A, Bonomo RP, Chillemi R, Sciuto S. Oxidation of 3-hydroxykynurenine. An EPR investigation. Journal of Heterocyclic Chemistry. 1990;**27**:2207-2208. DOI: 10.1002/jhet.5570270762

[51] Insausti TC, Le Gall M, Lazzari CR. Oxidative stress, photodamage and the role of screening pigments in insect eyes. The Journal of Experimental Biology. 2013;**216**:3200-3207. DOI: 10.1242/jeb.082818

[52] Ostrovsky MA, Dontsov AE. Vertebrate Eye Melanosomes and Invertebrate Eye Ommochromes as Antioxidant Cell Organelles: Part 2. The Biological Bulletin. 2019;**46**:105-116. DOI: 10.1134/S1062359019010084

[53] Lugo-Huitrón R, Blanco-Ayala T, Ugalde-Muñiz P, Carrillo-Mora P, Pedraza-Chaverrí J, Silva-Adaya D, et al. On the antioxidant properties of kynurenic acid: free radical scavenging activity and inhibition of oxidative stress. Neurotoxicology and Teratology. 2011;**33**:538-547. DOI: 10.1016/j.ntt.2011.07.002

[54] Romero Y, Martinez A. Antiradical capacity of ommochromes. Journal of Molecular Modeling. 2015;**21**:220. DOI: 10.1007/s00894-015-2773-3

[55] Lapina VA, Dontsov AE, Ostrovsky MA, Emanuél' NM. Interaction of oxygen anion radicals with eye melanins and ommochromes. Doklady Akademii Nauk SSSR. 1985;**280**:1463-1465. PMID: 2985349. (Article in Russian)

[56] Ostrovsky MA, Fedorovich IB, Dontsov AE. Photo-oxidative processes in eye structures. Protective function of the lens and shielding pigments. Biophysics (Moscow). 1987;**32**:896-909. PMID: 3318942. (Article in Russian)